my **revisi⊕n** notes

OCR (A) A Level

MATHEMATICS (APPLIED)

Stella Dudzic

Rose Jewell

Series editor
Roger Porkess

HODDER
EDUCATION
AN HACHETTE UK COMPANY

Every effort has been made to trace all copyright holders, but if any have been inadvertently overlooked, the Publishers will be pleased to make the necessary arrangements at the first opportunity.

Although every effort has been made to ensure that website addresses are correct at time of going to press, Hodder Education cannot be held responsible for the content of any website mentioned in this book. It is sometimes possible to find a relocated web page by typing in the address of the home page for a website in the URL window of your browser.

Hachette UK's policy is to use papers that are natural, renewable and recyclable products and made from wood grown in sustainable forests. The logging and manufacturing processes are expected to conform to the environmental regulations of the country of origin.

Orders: please contact Bookpoint Ltd, 130 Park Drive, Milton Park, Abingdon, Oxon OX14 4SE. Telephone: +44 (0)1235 827827. Fax: +44 (0)1235 400401. Email education@bookpoint.co.uk. Lines are open from 9 a.m. to 5 p.m., Monday to Saturday, with a 24-hour message answering service. You can also order through our website: www.hoddereducation.co.uk

ISBN: 978 1 5104 1763 2

First published in 2018 by
Hodder Education,
An Hachette UK Company
Carmelite House
50 Victoria Embankment
London EC4Y 0DZ

www.hoddereducation.co.uk

Impression number   10 9 8 7 6 5 4 3 2 1

Year   2022 2021 2020 2019 2018

Cover photo © Getty Images/Thinkstock/iStockphoto

Typeset in Bembo Std Regular 11/13 by Integra Software Services Pvt. Ltd., Pondicherry, India

Printed in Spain

A catalogue record for this title is available from the British Library.

Get the most from this book

Welcome to your Revision Guide for the Applied (Statistics and Mechanics) content of the OCR (A) A Level in Mathematics course. This book will provide you with reminders of the knowledge and skills you will be expected to demonstrate in the exam with opportunities to check and practice those skills on exam-style questions. Additional hints and notes throughout help you to avoid common errors and provide a better understanding of what's needed in the exam.

In order to revise the Pure Mathematics content of the course, you will need to refer to My Revision Notes: OCR (A) A Level Mathematics (Pure). The material in this book also covers the OCR (A) AS Level Mathematics exam; however, students may prefer to use the Revision Guide for the Applied content of the OCR (A) Year 1/AS Level Mathematics course which just covers all the applied mathematics needed for that exam.

Included with the purchase of this book is valuable online material that provides full worked solutions to all the 'Target your revision', 'Exam-style questions' and 'Review questions', as well as full explanations and feedback to each answer option in the 'Test yourself' multiple choice questions. The **online material** is available at **www.hoddereducation.co.uk/ MRNOCRALApplied**.

Features to help you succeed

Target your revision

Use these questions at the start of the two sections (one on Statistics, the other on Mechanics) to focus your revision on the topics you find tricky. **Short answers** are at the back of the book, but use the **worked solutions online** to check each step in your solution.

About this topic

At the start of each chapter, this provides a concise overview of its content.

Before you start, remember

A summary of the key things you need to know **before** you start the chapter.

Key facts

Check you understand all the key facts in each subsection. These provide a useful checklist if you get stuck on a question.

Worked examples

Full worked examples show you what the examiner expects to see in order to ensure full marks in the exam. The examples cover a sample of the type of questions you can expect.

Hint

Expert tips are given throughout the book to help you do well in the exam.

Common mistakes

Your attention is drawn to typical mistakes students make, so you can avoid them.

Test yourself

Succinct sets of multiple-choice questions test your understanding of each topic. Check your **answers online**. The **online feedback** will explain any mistakes you made as well as common misconceptions, allowing you to try again.

Exam-style questions

For each topic, these provide typical questions you should expect to meet in the exam. **Short answers** are at the back of the book, and you can check your working using the **online worked solutions**.

Review questions

After you have completed each of the two sections in the book, answer these questions for more practice. **Short answers** are at the back of the book, but the **worked solutions online** allow you to check every line in your solution.

At the end of the book, you will find some useful information:

Exam preparation

Includes hints and tips on revising for the A Level Mathematics exam, and information about the structure of the exam papers.

Make sure you know these formulae for your exam ...

Provides a succinct list of all the formulae you need to remember and the formulae that will be given to you in the exam.

Please note that the formula sheet as provided by the exam board for the exam may be subject to change.

During your exam

Includes key words to watch out for, common mistakes to avoid and tips if you get stuck on a question.

My revision planner

Statistics

REVISED   TESTED   EXAM READY

REVISED   TESTED   EXAM READY

Go online for
- **full worked solutions and answers to the Test yourself questions**
- **full worked solutions to all Exam-style questions**
- **full worked solutions to all Review questions**
- **full worked solutions to the Target your revision**

www.hoddereducation.co.uk/MRNOCRALApplied

STATISTICS

1 Understand the terms population and sample

To understand how people's friendships change as they age, a researcher interviews the same 100 people aged 60 and over every year for 10 years. For this study, identify:

i the sample

ii the population.

(see pages 6–7)

2 Use simple random sampling

All phone numbers in a town start with the same two digits (65). These are followed by four more digits to make the phone number. A researcher generates a 4-digit random number and then rings 65 followed by the random number. He does this 100 times. Will this give a simple random sample of the population in the town? Justify your answer.

(see pages 6–7)

3 Understand sampling methods: opportunity sampling, systematic sampling, stratified sampling, quota sampling, cluster sampling, self-selected samples

A group of market researchers interview people in different towns to ask about their shopping habits. Each researcher is told to interview the following numbers of people from different groups.

Age group	16–20	21–30	31–45	46–65	66 and over
Male number	10	20	35	40	25
Female number	10	20	35	40	30

Which sampling method is this?

(see pages 6–7)

4 Recognise possible sources of bias when sampling

A researcher wants to investigate how many people have tried illegal drugs. She puts a link to an anonymous online questionnaire in a tweet and asks her twitter followers to pass the link on. Identify two possible sources of bias in her sampling method.

(see page 9)

5 Be aware of practicalities of implementation of sampling methods

A union leader wants to find out about working conditions for her union members. She chooses a random sample of 100 union members to send questionnaires to. She knows that some of the sample will not return the questionnaires. Which one of the following options would be the best course of action? Justify your answer.

A Give some questionnaires to local union members she knows to make up numbers.

B Work with the questionnaires that are returned and do not try to replace them.

C Use a larger random sample to allow for some questionnaires not being returned.

(see page 9)

6 Recognise categorical, discrete, continuous and ranked data

Which of the words categorical, discrete, continuous, ranked would apply to measurements of height?

(see page 13)

7 Interpret a bar chart

Employees of film companies in Great Britain

Show that the number of employees involved in production has risen by about 50% from 2009 to 2016.

(see pages 14–15)

8 Interpret a dot plot or vertical line chart

The dot plot shows the year of first registration for each car on sale at a second-hand car dealer.

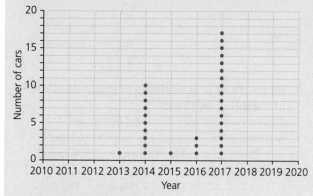

How many cars are on sale?

(see page 13)

9 Interpret a histogram or frequency chart

The histogram below shows CO_2 emissions for each of a random sample of 100 cars.

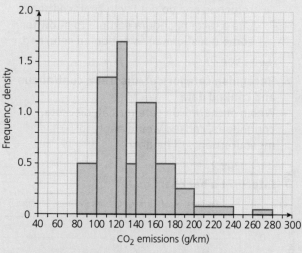

Describe the distribution of CO_2 emissions. Make two distinct comments about the information shown in the histogram.

(see pages 24–25)

10 Calculate frequencies or proportions from a histogram

For the histogram in Question 9, how many cars have CO_2 emissions between 120 and 130 g/km?

(see pages 24–27)

11 Interpret a pie chart

The pie charts below show the proportions of electricity generated in the US from different sources in January 2001 and December 2017.

US electricity generation
Jan 2001

US electricity generation
Dec 2017

□ Coal
□ Natural gas
▣ Conventional hydroelectric
■ Petroleum liquids
□ Nuclear
▣ Other renewables

Make two comments about changes in sources for electricity generation.

(see pages 13–14)

12 Interpret a stem-and-leaf diagram, use and interpret median and quartiles

The following stem-and-leaf diagram shows the hand lengths of a sample of 28 boys aged 10.

13	2 8
14	2 2 4 4 5 6 6 7 7 8 9 9
15	0 2 3 4 5 6 6 8 9 9
16	0 0 0 5

Key 13|2 means 13.2 cm

Find:
i   the median
ii the interquartile range.

(see page 16)

13 Interpret box plots

The box plots below show the CO_2 emission in g/km for three types of car.

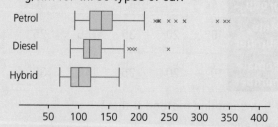

Compare CO_2 emissions for the three types of car. Make two distinct comments about the information shown by the box plots.

(see page 35)

14 Interpret a cumulative frequency diagram

1000 students sit an examination which is marked out of 100. Their results are shown in the cumulative frequency diagram below.

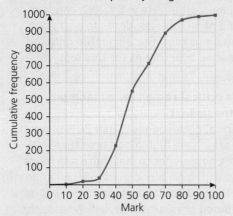

How many students got over 55 marks.

(see page 38)

15 Select or critique data presentation techniques

The chart below shows the gender of mathematics trainee teachers in England. State two ways to improve the presentation of the data.

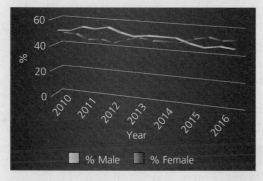

(see pages 13–14)

16 Calculate and interpret measures of central tendency: median, mode, mean

The age distribution of classroom teachers in state schools in England in 2016 is shown in the following table.

Age	Frequency (thousands)
Under 25	30.8
25–29	83.6
30–34	78.0
35–39	66.7
40–44	55.9
45–49	46.9
50–54	38.1
55–59	24.8
60 and over	11.7

Calculate an estimate of the mean age of classroom teachers. State any assumptions you make.

(see pages 21–22)

17 Calculate and interpret variance and standard deviation

For the data in Question 16, calculate an estimate of the standard deviation.

(see pages 43–44)

18 Identify outliers and clean data

i For the data in Question 16, show that there are some ages which could be classified as high outliers.

ii Are these values the result of errors? Justify your answer.

(see page 44)

19 Calculate probability for independent events

i For a particular type of chocolate bar, one bar in every 10 has a voucher for a free bar inside the pack. Jason buys 5 of these bars. What is the probability that he does not get a voucher for a free bar? Assume that vouchers occur independently of each other.

ii Is the assumption of independence reasonable?

(see pages 48 and 51)

20 Calculate probability for mutually exclusive events

A player in a game throws two fair dice. What is the probability of getting exactly one six?

(see page 50)

21 Use Venn diagrams to assist in calculation of probabilities

Events A and B are such that $P(A) = 0.8$, $P(A \cap B) = 0.5$ and $P(A \cup B) = 0.9$. Find $P(B)$.

(see pages 48–50)

22 Decide whether two events are independent

For the events A and B in Question 21, determine whether they are independent.

(see page 56)

23 Calculate conditional probability using the formulae

For events S and T, $P(S) = 0.4$, $P(T) = 0.3$ and $P(S \cap T) = 0.2$.
Find $P(S \mid T)$.

(see pages 54–55)

24 Calculate conditional probability using a tree diagram, Venn diagram or sample space diagram

All patients at a local health centre were asked whether they had made any of the following three changes during the past year.
- Eat healthier food.
- Increase exercise.
- Drink less alcohol.

The results of the survey are shown in the Venn diagram.

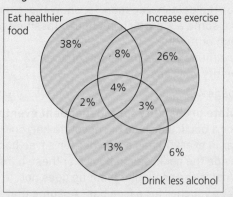

Given that a patient at the health centre has increased exercise, find the probability that the patient has drunk less alcohol.

(see pages 54–56)

25 Recognise situations which give rise to a binomial distribution

Could the number of rainy days in a week be modelled by a binomial distribution? A rainy day is defined as a day when any rain falls. Justify your answer.

(see pages 61–62)

26 Calculate probabilities for a binomial distribution

$X \sim \text{B}(48, 0.3)$. Calculate $\text{P}(X > 20)$.

(see pages 63–64)

27 Calculate expected value and expected frequencies for a binomial distribution

A biased coin has a probability 0.46 of landing heads. It is tossed 20 times. What is the expected number of heads?

(see pages 63–64)

28 Understand terminology associated with hypothesis testing

Neil is conducting a statistical hypothesis test. The p-value is 0.03. Neil says this means there is only a 3% chance of the null hypothesis being true. Is Neil right? Justify your answer.

(see page 70)

29 Know when to apply 1-tail or 2-tail tests when setting up a binomial hypothesis test

A website claims that half of men don't like football. Sadiq thinks that in his college most men like football. He will take a random sample of 50 male students to carry out a hypothesis test. State suitable null and alternative hypotheses for this test.

(see pages 71 and 77)

30 Find a critical region for a binomial hypothesis test

Historical data suggests that 30% of adults in an area have high blood pressure. A researcher suspects that there are more adults with high blood pressure. A random sample of 60 adults from the area will have their blood pressure checked. What is the critical region for a suitable hypothesis test at the 5% level of significance?

(see pages 74–75)

31 Conduct a hypothesis test using the binomial distribution and interpret the results in context

The researcher from Question 30 finds that the number of adults with high blood pressure in the sample is in the critical region. What should she conclude?

(see page 73)

32 Calculate probabilities for a simple discrete random variable

A discrete random variable can take the values 1, 2, 3 or 4 with probabilities as shown in the table below.

x	1	2	3	4
$\text{P}(X = x)$	0.3	0.3	0.2	k

i   Find the value of k.
ii   Find $\text{P}(X > 1)$.

(see pages 83–85)

33 Use the Normal distribution as a model

The histogram shows the upper arm lengths, in centimetres, of a sample of American adults.

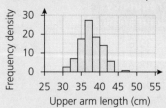

Does the shape of the histogram suggest that a Normal model is suitable? Justify your answer.

(see page 95)

34 Calculate and use probabilities from a Normal distribution

Upper arm lengths of adults are Normally distributed with mean 37.4 cm and standard deviation 2.8 cm. Find the probability that a randomly chosen adult has upper arm length greater than 45 cm.

(see pages 89–90)

35 Know the sampling distribution of the sample mean from a Normal distribution

Upper arm lengths of adults are Normally distributed with mean 37.4 cm and standard deviation 2.8 cm. What is the distribution of means of random samples of 10 adult upper arm lengths?

(see page 100)

36 Identify suitable null and alternative hypotheses for a Normal hypothesis test

Reaction times for adults are Normally distributed with mean 0.2 seconds and standard deviation 0.04 seconds. A researcher suspects that tired adults will have longer reaction times. She will find the reaction times of a random sample of 20 tired adults and conduct a suitable hypothesis test. What should be the null and alternative hypotheses for the test?

(see page 99)

37 Find a p-value for a Normal hypothesis test

The researcher conducting the hypothesis test in Question 36 finds that the sample mean reaction time is 0.23 seconds. Find the p-value. Assume the standard deviation remains 0.04 seconds.

(see page 104)

38 Interpret a scatter diagram and best fit model, describe correlation

Measurements of pulse rate and blood pressure are used to monitor people's health. Measuring pulse rate does not need special equipment so it can be done by anyone. A student is investigating whether measuring pulse rate also gives information about blood pressure. He takes a random sample of adults and draws the following scatter diagrams for systolic blood pressure, in mm Hg, against pulse rate, in beats per minute.

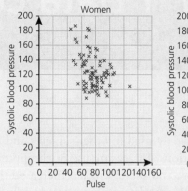
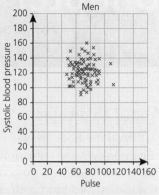

The correlation coefficient for women is −0.375 and for men is 0.005.

The student says that systolic blood pressure could be calculated from pulse rate for women but not for men. Is the student correct? Justify your answer.

(see pages 109–110)

39 Recognise outliers and different groups in scatter diagrams

The scatter diagram shows the number of teachers and the pupil–teacher ratio in all the schools in one area of the country. The correlation coefficient is −0.06.

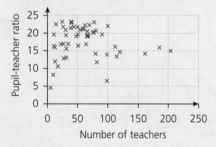

Make two distinct comments about what the scatter diagram shows.

(see page 110)

40 Use a given correlation coefficient to make an inference about correlation or association

In a triathlon event, athletes swim, cycle and run. A sports scientist suspects there is positive correlation between the cycling times and the running times so he conducts a hypothesis test using a random sample of 20 athletes. The correlation coefficient is 0.52; the 1-tail p-value is 0.0094 and the 2-tail p-value is 0.0188. What conclusion should the sports scientist draw from the test at the 1% level of significance?

(see page 113)

Short answers on pages 221–222

Full worked solutions online

CHECKED ANSWERS

Chapter 1 Data collection

About this topic

Statistical methods are widely used in other subjects and in the workplace to gain insight into situations. Increasing use of computers has made data easier to collect, store and share; this has increased the potential uses of statistics. Working with real data will involve deciding which of the statistical techniques you know are appropriate.

Before you start, remember ...

● Knowing what the mode, median and mean of a data set are.
● Interpreting bar charts and pie charts.

Data collection

REVISED

Key facts

1   The **problem solving cycle** shows the steps you follow when solving a problem.

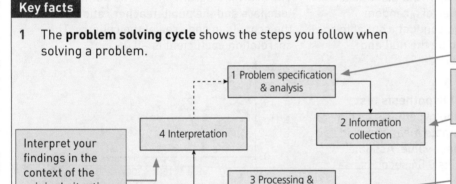

When working with statistics it is important to know what you want to achieve at the start so that you can collect relevant data.

This stage will usually involve taking a sample from the population.

Clean the data, present it in a suitable diagram and do any relevant calculations.

2   When **taking a sample**, you should ensure that:
   ● the data are **relevant**
   ● the data are **unbiased**
   ● the data are **not distorted** by the act of collection
   ● a **suitable person** is collecting the data
   ● the **sample** is of a suitable **size**
   ● a suitable **sampling procedure** is being followed.

Bias is a systematic error in the sampling.

3   Possible sampling methods include the following:
   ● In **simple random sampling**, every possible sample of a given size has an equal chance of being selected. A **sampling frame** is needed.
   ● In **stratified sampling**, the population is divided into sub-groups called **strata**. A sample is taken from each of the strata. If the samples from the strata are proportional to the size of their populations then this is called **proportional stratified sampling**.
   ● In **cluster sampling**, the population is divided into groups called clusters which are likely to be reasonably representative of the whole population. The sample is taken from a small number of clusters.

A sampling frame is a list (or equivalent) of the whole population. It is not always possible to have a sampling frame for the whole population.

A stratum (one of the strata) could be males or people in an age group – people in each stratum are unlikely to be representative of the population as a whole.

The sampling in each stratum is often done by simple random sampling.

A cluster could be people who live in the same town.

- In **systematic sampling**, the **sampling frame** is ordered and then individuals are chosen at regular intervals.

The first individual in a systematic sample is often chosen at random.

The sampling frame could be a list of the population on a spreadsheet.

The choice of who is sampled to fulfil the quota is usually left to the interviewer doing the survey.

- In **quota sampling**, the number to be sampled from each **stratum** of the population is decided. This method is often used for surveys.
- **Opportunity sampling** is used when a sample is readily available.
- A **self-selected sample** consists of volunteers.

For example, all the people at a meeting.

4 **Cleaning data** includes recognising and dealing with errors, outliers and missing data.

An **outlier** is a data item which is not consistent with the rest of the data. There are some rules for checking whether a data item is an outlier but sometimes it is obvious.

Worked example

Example 1

A researcher wants to investigate the reaction times of A Level Mathematics students.

The list below gives six things which may be relevant to the investigation.

Questionnaire	Measurement	A Level Mathematics students in England
A sample of 40 female and 60 male A Level Mathematics students chosen by the researcher at a university open day	A list of all A Level Mathematics students in England	A Level Mathematics students at a revision day

From the list, identify each of the following:

i the population

ii the sampling frame

iii an opportunity sample

iv a quota sample

v a possible method of collecting relevant data

The researcher would need to decide whether he/she wanted to investigate students in a particular year or over time.

Solution

i The population is A Level Mathematics students in England.

ii The sampling frame is a list of all A Level Mathematics students in England.

iii A Level Mathematics students at a revision day is an opportunity sample.

Common mistake: The students in this sample can all be easily tested at once – the students at the open day could not, so the revision day is the opportunity sample.

iv A sample of 40 female and 60 male A Level Mathematics students chosen by the researcher at a university open day is a quota sample.

v Measurement is a possible method of collecting relevant data.

Common mistake: A questionnaire can be a way of collecting relevant data but not in this case – students will not know their reaction times.

OCR (A) A Level Mathematics (Applied)

Worked example

Example 2

The dot plot below shows the number of mm of rainfall collected by a rain gauge in Leeds on each day in July 2013.

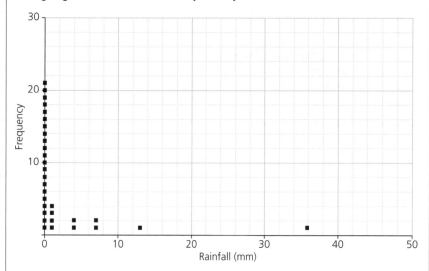

i State the mode.

ii Find the median.

iii Identify an outlier.

iv Is the outlier necessarily a mistake in the data? Explain your reasoning.

v The data were collected as part of an investigation into whether July in Leeds is getting wetter than it used to be. What other data would be needed?

Solution

i The mode is zero.

ii The median is zero.

iii The value which is about 35 mm is an outlier.

iv No, the outlier need not be a mistake. It is possible to have very rainy days in July.

v Rainfall data for Leeds for July for other years.

> Most of the days had no rain (over 20 days) so this is the mode.

> There are 31 days in July. If the amounts of rainfall were put in order, the 16th number would be the median. There are over 20 days with no rain so the 16th number must be zero.

> The value which is about 13 mm could also be considered an outlier.

> Outliers can be mistakes in the data but they need not be. Sometimes a value might be so high that it is clearly not possible but that is not the case here.

> When choosing a sample, you need to know what population you are selecting from. In this case, it is rainfall in Leeds in July over time.

Advantages and drawbacks of sampling methods

Method	Advantages	Drawbacks
Simple random sampling	• Avoids bias • Probability can be used to study properties of samples	• Needs a sampling frame • May be expensive or time-consuming to collect the sample
Stratified sampling	• Ensures all subgroups in the population are represented	• May be expensive or time-consuming to collect the sample • It may be hard to find out information about which subgroups individuals are in
Cluster sampling	• Does not need a sampling frame • Reduces cost and time spent in sample collection	• It's hard to know whether the sample is representative of the whole population or whether it is biased
Systematic sampling	• Easy and quick when a sampling frame is available	• Needs a sampling frame • Could be biased due to a pattern in the sampling frame
Quota sampling	• Ensures all specified subgroups in the population are represented	• It can be biased depending on how the individuals are chosen
Opportunity sampling	• Quick and easy	• The sample may not represent the population
Self-selected sampling	• Quick and easy	• Very likely to be biased

Worked example

Example 3

A journalist on the Avonford Star wants to know what the public thinks of a proposal to build a new housing estate in Avonford, so he invites people to add comments to the Avonford Star website.

i   What type of sample is this?

ii Give two reasons why the sample may be biased.

Solution

i   This is a self-selected sample.

ii Possible reasons include the following:

   • The sample excludes people who do not read the Avonford Star.

   • The sample excludes people who do not use the internet.

   • People may be influenced by seeing other people's comments.

   • Only people with strong views are likely to respond.

Hint: The people who respond decide to do so for themselves, so they are choosing to be in the sample rather than being chosen by someone else. This makes it self-selected.

Hint: Notice that there are four possible reasons given here. The question only asks for two reasons. Any two sensible reasons will do. You should not give more than two reasons. When asked to do something in an examination you need to follow instructions carefully.

Test yourself

1 A researcher investigating health has a sample of data about adult heights. The data give the height and sex of each person but individuals cannot be identified. The height of one man in the sample is listed as 6 metres. What should she do with this data value?

  A She should include it – it would bias the sample to ignore it.

  B She should delete it from the data set – it is clearly wrong.

  C She should not use the sample – it contains errors.

  D She should go and measure the height of the man herself.

  E She should realise that it is meant to be 6 feet and use that value for the height.

2 A student is investigating how students in his college use their leisure time. He starts with a list of all students at the college and looks at male and female students separately. He chooses equal numbers of male students and female students at random. Which sampling method is the student using?

  A Cluster sampling

  B Quota sampling

  C Simple random sampling

  D Stratified sampling

  E Systematic sampling

3 A biologist is investigating how many eggs a particular species of birds lays. She visits three sites where she knows the birds nest and counts the number of eggs in a random sample of nests at each site. What sampling method is the biologist using?

  A Stratified sampling

  B Simple random sampling

  C Quota sampling

  D Opportunity sampling

  E Cluster sampling

4 Which of the following is NOT true of simple random sampling?

  A Simple random sampling avoids bias.

  B A sampling frame is needed for simple random sampling.

  C A simple random sample will always be more representative of the population than any other type of sample.

  D Probability methods can be used to study properties of random samples.

  E If individuals chosen as part of a simple random sample refuse to take part then this may introduce bias.

5 A market researcher wants to know what users of a particular post office think of the facilities available. Which of the following is likely to be the least biased sampling method?

  A Choose everyone who is in the post office at noon on Monday.

  B Put an advert in the local paper and ask volunteers to get in touch if they use the post office and want to give their opinions about the facilities.

  C Visit the post office at twelve times during one week, two on each day Monday to Saturday, and choose every fifth person who comes into the post office during half an hour.

  D Ask post office staff to give a letter to 20 men and 20 women who use the post office. The letter asks the person who receives it to complete a questionnaire and give it back to the post office staff.

  E Put up a sign that can be seen by everyone entering the post office asking them to get in touch with the market researcher to give their opinion about the facilities. Choose a simple random sample from those who get in touch.

Full worked solutions online

Exam-style question

A market researcher wants to know whether customers of a particular supermarket can taste the difference between the supermarket own brand digestive biscuits and a well-known brand.

i   What is the population for this investigation?

ii Explain why it is not possible to take a simple random sample from the population.

The market researcher will visit one branch of the supermarket one Monday morning and ask 20 customers to taste both types of biscuit. She will choose 10 males and 10 females.

iii State the sampling method being used.

iv Suggest one possible improvement to the sampling method to get more reliable results from the investigation.

Short answers on page 222

Full worked solutions online

CHECKED ANSWERS

Chapter 2 Data processing, presentation and interpretation

About this topic

Statistics is about making sense of data; knowing what kind of data you have is important so that you can use appropriate techniques to help you unlock the information they contain. This chapter deals with statistical graphs and summary statistics.

Being able to understand statistical graphs and use them appropriately is an important skill. You are likely to encounter such graphs in the news, in other subjects and in your future work. Software, for example a spreadsheet, is often used to produce statistical graphs but care is needed to ensure that the correct type of graph is used to display the data and that the axes and labels are correct to ensure that a misleading impression is not given.

Averages are often used in everyday life, so you need to understand them thoroughly. Averages are sometimes called 'measures of central tendency'. Averages are particularly useful for summarising and comparing larger sets of data; these are usually presented as frequency tables. In addition to knowing the average, it is also important to know how spread out the data are. The interquartile range and standard deviation are commonly used measures of spread.

Histograms are usually used for continuous data which have been grouped; understanding them will give you a useful way of displaying data. You may already have studied them at GCSE but do check your understanding as it is quite easy to go wrong with histograms.

Cumulative frequency curves are often used to estimate the median, quartiles and percentiles of grouped data; these measures are used to compare individuals or groups of people. For example, growth curves for young children use them. Box-and-whisker diagrams use the median and quartiles to provide a visual way of comparing sets of data.

Scatter diagrams show relationships between two variables; considering how two variables are related to each other is important in statistical modelling to make predictions.

Before you start, remember ...

- How to draw bar charts and pie charts from GCSE.
- Putting numbers in order (including decimals and negative numbers) from GCSE.
- Frequency tables, rounding and cumulative frequency from GCSE.

Interpreting graphs for one variable

> ## Key facts
>
> 1 Categorical or qualitative data are not numerical.
> 2 Numerical or quantitative data can be subdivided into discrete or continuous.
> 3 The data may also be the ranks of the variables in the set.
> 4 Continuous data may take any values; these are usually within a range.
> 5 Discrete data may only take certain separate values, for example whole numbers.
> 6 Vertical line charts and dot plots are often used for discrete numerical data. The vertical axis shows the frequency. The horizontal axis is a continuous scale.
> 7 Bar charts are often used for categorical data. There should be gaps between the bars.
> 8 A compound bar chart stacks the bars for several sets of data on top of each other so they can be compared.
> 9 A multiple bar chart puts the bars for several sets of data next to each other so they can be compared.
> 10 Pie charts illustrate the frequencies of sections of the data set in comparison with each other and the whole set.
> 11 You can use a stem-and-leaf diagram for discrete or continuous data.
> 12 A stem-and-leaf diagram shows the shape of the distribution as well as keeping the original data values.
> 13 A distribution with one peak is unimodal. If it has two peaks, it is bimodal; the peaks need not be the same height.

Types of data

Categorical or **qualitative** data are in categories; they are not numerical. Examples include eye colour, breed of dog and favourite type of book.

Numerical or **quantitative** data consists of numbers.

Ranked data are the ranks of the variables in a set rather than their actual size.

Continuous variables are measured on a scale, e.g. length, weight, temperature. For any two possible values, you can always find another possible value between them. In practice, continuous data are always rounded because there is a limit to how accurately they can be measured.

Discrete variables are often just whole numbers, e.g. the number of children in a family, but they need not be, e.g. shoe sizes sometimes include half sizes.

> **Common mistake**: Sometimes categorical data are coded using numbers. For example, the method someone uses to travel to work might be coded using 01 for car, 02 for train, etc. The number is just a shorthand, it does not make the data numerical.

Using statistical diagrams

Statistical diagrams can give a quick visual impression of a whole set of data. It is important to use the right kind of diagram and to draw it correctly so that the impression given is an accurate one. The following flow chart will help you to understand when different graphs and charts are appropriate for discrete and continuous numerical data.

Start in the grey box towards the centre to help you make a decision about the best diagram to use. Sometimes there is more than one possibility.

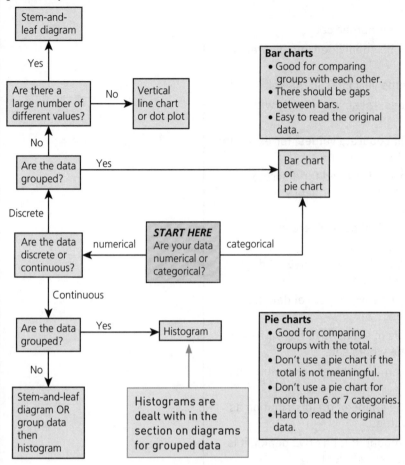

Compound bar charts and multiple bar charts

Compound bar charts and multiple bar charts are used to compare two, or more, sets of similar data.

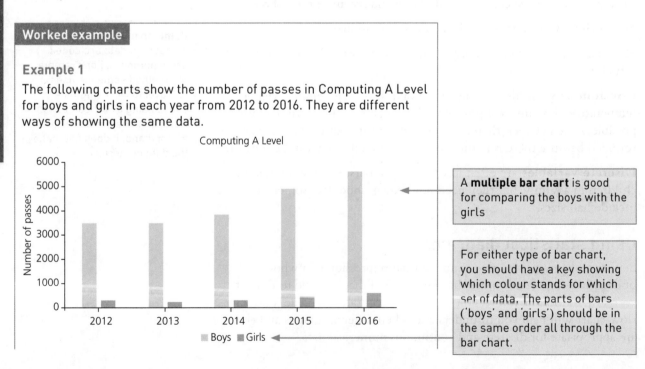

Worked example

Example 1

The following charts show the number of passes in Computing A Level for boys and girls in each year from 2012 to 2016. They are different ways of showing the same data.

A **multiple bar chart** is good for comparing the boys with the girls

For either type of bar chart, you should have a key showing which colour stands for which set of data. The parts of bars ('boys' and 'girls') should be in the same order all through the bar chart.

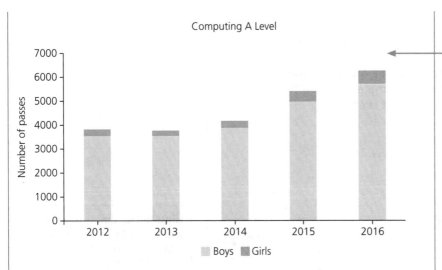

Computing A Level

A **compound bar chart** is good for comparing the total number of passes. You should not use this type of bar chart if the totals are not meaningful.

i   Describe the trend in Computing A Level passes.
ii The graphs do not show the data for 2017. Jack says that the graphs show that the total of Computing A Level passes in 2017 must have been greater than in 2016. Comment on Jack's statement.

Hint: Each graph shows individual passes for boys and for girls as well as total passes – your answer should say which numbers you are referring to. Since the question does not specify which number to look at, it is helpful to refer briefly to all the variables.

Solution

i   *The general trend is for the total number of passes in Computing to increase over time. Passes for each of boys and girls are also increasing over time.*

ii *Although the trend for total passes is consistently upwards over the timescale shown by the graphs, there is no guarantee that this trend will continue.*

Hint: You are asked to comment on the answer so say why Jack might have said what he did and also say whether this is correct. The trend in the graphs does not guarantee that the next year will continue to follow the trend.

Shapes of distributions

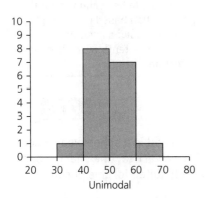

Unimodal

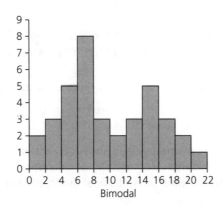

Bimodal

The value of the variable at the peak of a distribution shown is the mode. A **unimodal** distribution has one mode; a **bimodal** distribution has two modes (if they are not next to each other, one of them can have a higher frequency than the other).

Stem-and-leaf diagrams

- An ordered stem-and-leaf diagrams is drawn with the 'leaves' in order; this allows the median to be found easily from the diagram.
- For sets of data which have a lot of leaves on each row, two separate rows can be used for each 'stem' one for leaves 0 to 4 and the second for 5 to 9. Sometimes an asterisk is used on the first of the two stems.
- Back to back stem-and-leaf diagrams are useful for comparing two sets of data, as shown in the following example.

Worked example

Example 2

The back to back stem-and-leaf diagram below shows the pulse rates of two samples of people; those in one sample are aged 10–19 and those in the other sample are aged 70 and over.

Age 10–19		Age 70+
8	5	4 4 4 6 6 6 8 8
8 8 8 6 6	6	0 0 2 2 4 4 6 6 6 8 8 8
8 6 6 4 4 4 4 2 2 2 0 0 0	7	0 0 0 0 0 8
8 8 6 6 4 2 2 0	8	0 2 8
6 4	9	1
0	10	

Key: 4|9|1 represents 94 beats per minute for age 10–19; 91 beats per minute for age 70+

> **Hint:** Notice that you need to look at the key to see what the numbers mean.

> **Hint:** Notice that the leaves are ordered with the smallest near the stem.

i   Write down the highest pulse rate and state the age group it comes from.

ii Compare the shapes of the two distributions from this sample.

iii Explain whether the data show that pulse rates for older people in the population must be lower than for children.

Solution

i   The highest pulse rate is 100 beats per minute and is from the age 10–19 sample.

ii The pulse rates of the 10–19 age groups are slightly more spread out than for the age 70+ group. Both distributions are unimodal; the modal group for age 10–19 is 70 beats per minute but for age 70+ it is 60 beats per minute.

iii The pulse rates are generally lower for the older people in the sample but we cannot be certain whether this is also true for the population.

> **Common mistake:** It is never possible to be certain about the population based on a sample so conclusions in statistics need to be tentative rather than certain.

Test yourself

1 A head of year wants to illustrate how many days absence each of six classes had in a school year. Each class has the same number of students. He needs to decide whether to draw a vertical line chart, a bar chart or a pie chart. Which of these kinds of statistical diagram would be well suited for this purpose?

Form	Days absence
10A	217
10B	74
10F	99

A   Only a vertical line chart.

B   Only a bar chart.

C   Only a pie chart.

D   Either a vertical line chart or a bar chart but not a pie chart.

E   Either a bar chart or a pie chart but not a vertical line chart.

2 The pie chart illustrates the number of bedrooms in houses for sale advertised in the Avonford Star. There were 72 houses for sale altogether. How many had one bedroom?

A   1.8   B   2   C   10

D   50   E   Not possible to tell.

No. of bedrooms

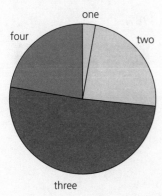

3 Four of the following statements about the graph below are false and one is true. Which one is true?

A   UK had 30 bronze medals.
B   Greece had more silver medals than gold medals.
C   France had more silver medals than the UK.
D   Germany had more silver medals than gold medals.
E   Ranking these countries in order of their number of gold medals gives the same result as ranking them by their total number of medals.

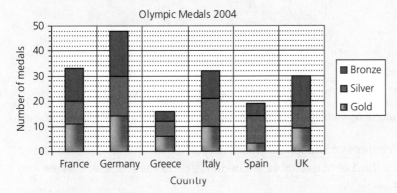

4 The cloudiness of the sky is measured in oktas; the number of oktas is the fraction of the sky (in eighths) covered by cloud. The vertical line charts below show the frequencies of the different possible values for the cloud cover for the days from May to October in 1987 (left) and 2015 (right) for Heathrow.

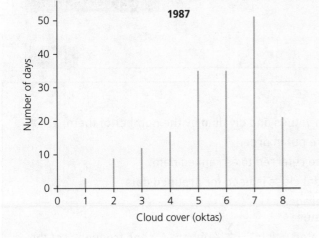 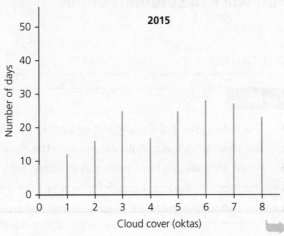

Four of the following statements are true and one is false. Which one is false?

A   The graphs show that the weather at Heathrow is getting gradually sunnier as time goes on.

B   There were similar numbers of days in May to October in each of 1987 and 2015 with completely cloudy sky.

C   In 1987, over one-quarter of the days for which data are shown had 7 okta cloud cover.

D   These data show that in both 1987 and 2015, more than half the sky was cloud covered on more than half of the days.

E   May to October 2015 was generally less cloudy than the same period in 1987.

Full worked solutions online

CHECKED ANSWERS

Exam-style question

The two charts below have been drawn using a spreadsheet. They show the percentage of household waste recycled in three regions of England.

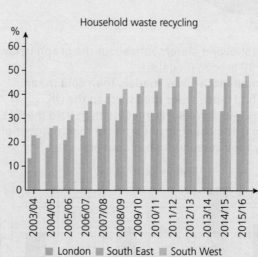

i   One of the charts is not an appropriate way to display the data. State which one and why it is not appropriate.

ii Make two statements about the percentage of household waste recycled over the time period shown.

iii A newspaper reports that less than a third of households in London recycle their waste. Explain why this statement is unlikely to be true.

Short answers on page 222

Full worked solutions online

CHECKED ANSWERS

Averages

REVISED

Key facts

1   The mean (symbol $\bar{x}$) is found by adding the data values and dividing by the number of them; $\bar{x} = \frac{\sum x}{n}$.

2   The median is the middle value when the data are put in order.

3   When data values have been put in order, they are referred to as ranked data.

4   The mode is the most common item of data (modal class is used for grouped data).

5   The range is a measure of spread and not an average.
    range = maximum data value − minimum data value

6   For data in a frequency table, the mean is calculated using $\bar{x} = \frac{\sum xf}{n}$ where f is the frequency of the data value x.

7   For a grouped frequency table, you can only calculate an estimate of the mean as you do not know the exact data values. The midpoint of each group is used when doing this.

8   The mode is the item with the highest frequency in an ungrouped frequency table.

9   The modal class is the group with the highest frequency in a grouped frequency table (if it has equal width groups).

Measures of central tendency (averages)

A measure of central tendency is a single value that is used to represent a set of data. They are sometimes called averages and are useful for comparing sets of data, e.g. teachers might compare classes of students by looking at the average test mark for each class.

Mean

The mean is often referred to as 'the average' in everyday speech. It is worked out by adding up the data values and dividing by the number of them. This can be written as the formula:

$$\bar{x} = \frac{\sum_{i=1}^{n} x_i}{n}$$

where

- $\bar{x}$ stands for the mean.
- The data items are $x_1, x_2 \ldots x_n$; x_i is a typical, or general, data item.
- The Greek letter sigma (Σ) means 'the sum of'. It gives the total of all the data.
- The number of data items is n.

The formula is often written more simply as:

$$\bar{x} = \frac{\sum x}{n}.$$

The mean is a 'fair shares' average. For example, working out the mean earnings of a group of people is the same as you would get working out how much each person would receive if they all shared their earnings equally among themselves.

You will usually find the mean by using statistical functions on a calculator but you need to know how it is worked out so that you can use the mean to calculate the total.

Median

The median is the middle value when the data items are put in order. If there is an even number of data values, there will be two middle values; the median is half way between them.

You are expected to use the statistical functions of your calculator to get the median – make sure you know how to do this. You should round your final answer sensibly. Most calculators will give a list of statistics once you have input the data.

Example 1

There are 20 students in a class; 19 of them measure their pulse rates and find the mean is 68. The 20th student measures his pulse rate; it is 97. What is the mean pulse rate for all 20 students?

Solution

The total for the 19 students is $19 \times 68 = 1292$

The total for all 20 students is $1292 + 97 = 1389$.

The mean for all 20 is $1389 \div 20 = 69.75$

> mean $= \dfrac{\text{total}}{n}$ so total $= n \times$ mean

> Even if all the data are whole numbers, the mean need not be.

Mode

The mode is the most commonly occurring value. There can be more than one mode, or there might not be a mode. The mode can be used for categorical data; the other measures of central tendency cannot.

> **Hint:** All the measures of central tendency produce a value that represents the whole set of data. If you get an answer for an average of numerical data that is either larger than the largest value in the set or smaller than the smallest value, you know you have made a mistake somewhere.

Deciding which average to use

- The mean is affected by one very large (or small) data value.
- The mean allows you to work out the total of the data and uses all the values.
- The median is not affected by extreme data values.
- The mode gives the most likely value to occur; it can be used for categorical data.
- For symmetrical, unimodal data the mean, median and mode are all equal.

Averages from an ungrouped frequency table

Example 2

The table below shows the number of goals scored by the winning team (or the goals scored by either team in the case of a draw) for a sample of league football matches.

Number of goals	0	1	2	3	4	5
Frequency	12	7	11	4	0	1

> If a data value has frequency zero, this means it did not occur. There were no games where the winning team scored 4 goals.

> **Common mistake:** In a frequency table for numerical data, there are two sets of numbers. The frequency tells you how often each value occurred; the other numbers are the data values that you are dealing with. Understanding this will help you avoid some of the common mistakes made when working with frequency tables.

i   Find the mean.
ii Find the median.
iii Explain how you can tell from the data that at least 12 games were draws.

	List 1	List 2	List 3	List 4
SUB				
1	0	12		
2	1	7		
3	2	11		
4	3	4		

Hint: Your calculator should allow you to type in the data as a frequency table. Make sure you know how to do this for your calculator.

Solution

i   The mean is 1.31 (to 2 decimal places).

	1–Variable
$\bar{x}$	1.31428571
Σx	46
Σx^2	112
σx	1.21352917
sx	1.23124586
n	35

Make sure you write down the correct value from the list of statistics in your calculator. You should also check that the value of n (the number of data values) is what you expect – in this case there were 35 games; this is the total frequency.

Common mistake: If your calculator uses lists make sure that the frequency is set to the correct list.

1 Var	XList	: List 1
1 Var	Freq	: List 2

ii The median is 1.

Most calculators will give the mean and the median as part of the same list of statistics

iii For the games with zero goals, if the winning team scored zero then the other team cannot have scored fewer goals so all 12 games with zero goals must be draws.

Hint: The question mentions 12 games, so it makes sense to think about the 12 games with zero goals; remember that the data are the number of goals scored by the winning team or both teams and it is not possible to score less than no goals.

For the data in Example 2, it is easy to see that the **mode** is zero goals, because this happened 12 times (the highest frequency).

Averages from a grouped frequency table

For the **mean**, you do not know all the individual data values so you use the midpoint for each group as an estimate of all the values in it; this means that your eventual answer is an estimate of the mean.

The table below shows the time of the first goal for a sample of league football matches, together with the midpoint of each group.

Hint: The mean can be found using a calculator in the same way as for an ungrouped table **except** that you are working with the midpoint of each group as an estimate of the values in that group. It is a good idea to write down the midpoints to help you type them into the calculator correctly.

Time (minutes)	Frequency, f	Midpoint, x
1–10	10	5.5
11–20	5	15.5
21–30	8	25.5
31–45	8	38.0
46–60	4	53.0
61–70	3	65.5
71–80	3	75.5
81–90	2	85.5

Using the midpoint of the group as an estimate for all the data values in the group loses information; this does not matter too much for the estimation of the mean as that is based on the total of all data values.

Common mistake: The times have been rounded before being put into the frequency table. The 11–20 minute group is really 10½ ≤ time < 20½; in this case the midpoint would be the same so it does not make a difference. Be especially careful when working with ages; they are always rounded down. The 11–20 years group would really be 11 ≤ age < 21 years, with a midpoint of 16.0.

Common mistake: It would not make sense to state the **modal class** for this table because there are some groups wider than others; if they were the same width, things could look different.

The mean is 33.6 (to 3 s.f.).

Hint: It can be easy to go wrong when working out mean from a frequency table; always look back at the data to see whether your final answer is a reasonable average.

The **median** is estimated by using a cumulative frequency graph; this is revised in the next two sections in this chapter.

Common mistake: When you enter the midpoints and frequencies into your calculator, you are using the midpoint of each group as an estimate of all data values in that group; this does **NOT** give a good estimate for the median of grouped data. The same applies to the quartiles. Only the mean, standard deviation and various totals should be used from the list of statistics which the calculator gives.

Test yourself

1 A sample of students are asked how many school dinners they ate in the past week with the following results.

Number of dinners eaten	0	1	2	3	4	5
Frequency	7	10	10	12	21	18

Four of the statements below are false and one is true. Which one is true?

A The median is 3.5.

B The mode is 10.

C The number of students in the sample was 6.

D The number of students in the sample was 15.

E The median is 2.5.

2 A group of students were asked how many books they have in their room and gave answers as follows:
5 5 3 2 1 3 2 57 2 9 3 8
Four of the following statements are false and one is true. Which one is true?

A The mode is 2.5.

B The mean should not be used for these data because it is impossible to have 8.333... books.

C There is no limit to the number of books a student could have so the data are continuous.

D The median, which is 3, is a good average to use as it is unaffected by the large value of 57.

E The median is 2.5.

3 A class of students consists of 9 girls and 20 boys. The mean weight of the girls is 53 kg; the mean weight of the boys is 61 kg. What is the mean weight for the whole class?

A 3.9 kg   B 55.5 kg   C 57 kg   D 58.5 kg   E 848.5 kg

4 A teacher has to choose someone from a class to represent them in a regional spelling competition. They have regular spelling tests (marked out of 10) in the class and two students, who have taken all 20 of the class tests, are willing to represent the class. Information about their average scores in the class tests is shown below. Four of the following statements are true and one is false. Which one is false?

	Mark	Lucy
Mean	6.8	7.6
Mode	8	6

A   Because there were 20 class tests, each of them must have scored their mode more than twice.

B   Their means show that Mark scores less than 7 half the time and Lucy scores more than 7 half the time.

C   The distribution of Mark's scores cannot be symmetrical.

D   For a short competition, Mark is a better choice than Lucy as his higher mode shows that he is more likely to do well.

E   For a long competition, Lucy is a better choice than Mark as her higher mean shows that her total score is higher than his.

5 The waist measurements of a sample of boys are shown below.

Waist (cm)	50–59	60–69	70–79	80–89	90–109
Frequency	2	45	80	19	7

Four of the statements below are false and one is true. Which one is true?

A   The mode is 80 cm.

B   The mean is 30.6 cm.

C   A good estimate of the mean is 74.5 cm.

D   The median is 74.5 cm.

E   The median is somewhere in the group 70–79 cm.

Full worked solutions online

CHECKED ANSWERS ☐

Exam-style question

The table below shows the ages of the residents of the City of London at the time of the 2011 census.

Age	0–15	16–19	20–29	30–44	45–59	60–74	75–89	90 and over	Total
Frequency	620	164	1501	2045	1547	1050	406	42	7375

i   Alan is calculating an estimate of their mean age. He starts by saying that the midpoint of the first group is 7.5. Is this correct? Explain your answer.

ii   Britney correctly calculated an estimate of the mean age as 42.3 years (to 1 d.p.). To do this, she had to choose an upper age for the last group. Show that this must have been at least 94.

iii The National Statistics website reports the mean age as 41.4. Why is this different from the mean Britney calculated?

Short answers on page 222

Full worked solutions online

CHECKED ANSWERS ☐

Diagrams for grouped data

Key facts

1 **Histograms**
   - Histograms are usually used for continuous grouped data.
   - The horizontal axis on a histogram is an even scale showing the values of the variable (e.g. height, wage, age).
   - The vertical axis shows frequency density; this is calculated by dividing the frequency by the class width.
     - The label can be 'frequency density' or, for example, 'frequency per cm'.
     - The symbol / can be used instead of 'per'.
   - For a histogram with equal width bars, software often puts frequency on the vertical axis rather than frequency density; such a diagram is a frequency chart.
   - Because the horizontal axis on a histogram is a continuous scale, there are no gaps between bars.
   - The frequency represented by a bar in a histogram is given by the area of the bar.
   - The modal class is the group with the highest frequency density.

2 **Cumulative frequency graphs**
   - Points on a cumulative frequency graph are plotted at the upper boundary of the group.
   - Cumulative frequency tells you how many data items were up to a particular value.

Histograms for continuous data with unequal class widths

Worked example

Example 1

The maximum daily temperature in central England in February 2017 is given in the table below. Draw a histogram to illustrate these data.

Temperature (°C)	Frequency	Class width (°C)	Frequency density
$0 \leqslant T < 2$	1	2	0.5
$2 \leqslant T < 4$	2	2	1
$4 \leqslant T < 6$	3	2	1.5
$6 \leqslant T < 8$	3	2	1.5
$8 \leqslant T < 9$	3	1	3
$9 \leqslant T < 10$	5	1	5
$10 \leqslant T < 11$	4	1	4
$11 \leqslant T < 12$	5	1	5
$12 \leqslant T < 14$	1	2	0.5
$14 \leqslant T < 16$	1	2	0.5

Using narrower groups where there are more data (in the middle for this example) can help to show an appropriate level of detail

Hint: Frequency density is calculated by dividing the frequency by the class width; it is useful to have a column in the table for the class widths.

Solution

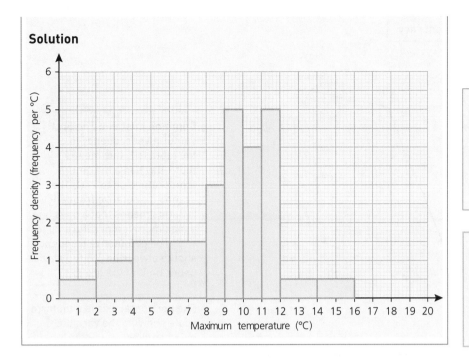

Hints:
- The area of a bar in a histogram is the frequency of the group it represents.
- The vertical axis could be labelled either 'frequency density' or 'frequency per °C'.

Common mistake: When drawing a histogram for **age**, remember that, for example, a group aged 14–19 years could have people that are just a day short of their 20th birthday so its width would be 20 – 14 = 6 years.

Cumulative frequency graphs

The maximum daily temperature in central England in February 2017 is given in the following table, together with the cumulative frequencies. These are the same data as for Example 1.

Temperature (°C)	Frequency	Cumulative frequency
$0 \leqslant T < 2$	1	1
$2 \leqslant T < 4$	2	3
$4 \leqslant T < 6$	3	6
$6 \leqslant T < 8$	3	9
$8 \leqslant T < 9$	3	12
$9 \leqslant T < 10$	5	17
$10 \leqslant T < 11$	4	21
$11 \leqslant T < 12$	5	26
$12 \leqslant T < 14$	1	27
$14 \leqslant T < 16$	1	28

The entries in the cumulative frequency column are found by adding up the frequencies as you go down the table. In this example the cumulative frequency tells you how many days were in the given maximum temperature group and the ones below it.

Hint: The final cumulative frequency gives the total frequency.

The cumulative frequency graph is shown on the next page.

The vertical axis shows cumulative frequency.

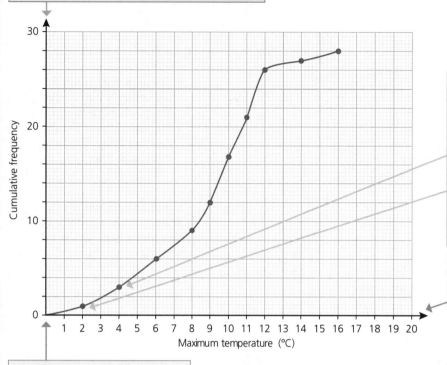

Common mistake: The first cumulative frequency in the table is 1; this means that 1 day had a maximum temperature up to 2 °C. The second cumulative frequency is 3; this means that 3 days had a maximum temperature up to 4 °C. The cumulative frequency should always be plotted at the upper limit of the group.

The horizontal axis is a uniform scale showing the variable being studied.

Hint: There were no days with a maximum temperature below 0 °C so the cumulative frequency for 0 °C is 0.

Worked example

Example 2

Use the cumulative frequency graph to estimate the number of days in February 2017 with a maximum temperature of over 5 °C.

Solution

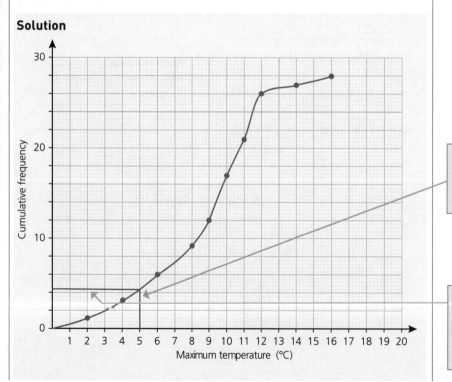

Draw a line up from the temperature of 5 °C until it hits the cumulative frequency graph.

Draw a line across from where the vertical line hits the graph to get the cumulative frequency that corresponds to a temperature of 5 °C.

The cumulative frequency for 5°C is between 4 and 5. This means that there were either 4 or 5 days with a maximum temperature of 5°C or less. There must be a whole number of days and 5 is too many so there were 4 days with a maximum temperature of 5°C or less The number with maximum temperature over 5°C is 28 − 4 = 24.

Hint: For high bars, the cumulative frequency goes up more quickly as you go along the bar so the cumulative frequency graph for that range is steeper.

The relationship between the cumulative frequency graph and the histogram

The histogram and the cumulative frequency graph for the data in Example 1 are shown side by side below.

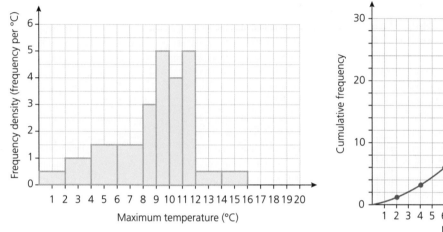
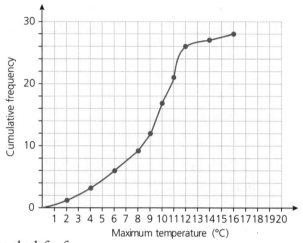

The area in the histogram represents frequency. The area to the left of a vertical line in the histogram is the cumulative frequency and so is the same as the height of the cumulative frequency curve for that value of the variable.

Test yourself

TESTED

1 A sample of students was asked to estimate the length of a line. The data are shown in the table. Four attempts at a histogram for these data are shown below. Three of them are incorrect and one is correct. Find the one which is correct.

Estimate (cm)	$14 \leqslant l < 18$	$18 \leqslant l < 22$	$22 \leqslant l < 24$	$24 \leqslant l < 26$	$26 \leqslant l < 30$	$30 \leqslant l < 34$	$34 \leqslant l < 42$
Frequency	12	7	13	7	8	7	6

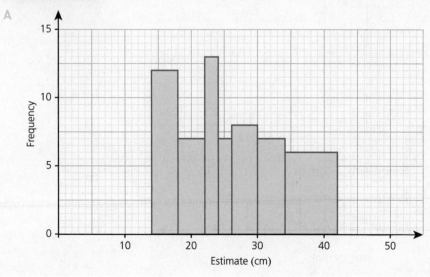

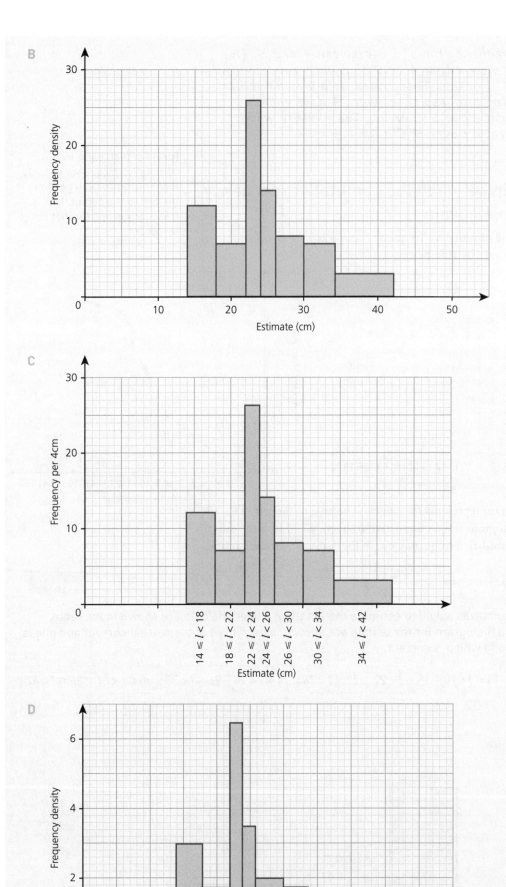

2 The following table shows the ages of a sample of the customers of a shop.

Age	13–16	17–19	20–25	26–35	36–55	56–70
Frequency	8	15	19	17	15	5

Between what limits should the next to last bar in the histogram be drawn?

A   35 to 55   B   36 to 56   C   35.5 to 55.5   D   35 to 56   E   36 to 55

3 The histogram below shows the pulse rates of a sample of people. How many people were in the sample?

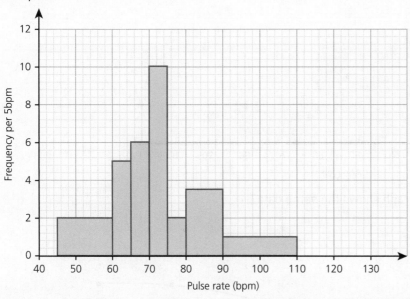

A   7   B   29.5   C   30   D   40   E   200

4 Which of the frequency tables below shows the data corresponding to this cumulative frequency curve?

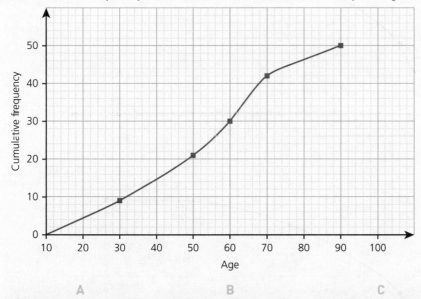

	A		B		C		D

Age	Frequency	Age	Frequency	Age	Frequency	Age	Frequency
10–30	9	10–29	9	20–40	9	11–30	9
31–50	21	30–49	12	40–55	12	31–50	21
51–60	30	50–59	9	55–65	9	51–60	30
61–70	42	60–69	12	65–80	12	61–70	42
71–90	50	70–89	8	80–100	8	71–90	50

5 The histogram below shows the age distribution of the population of a city.

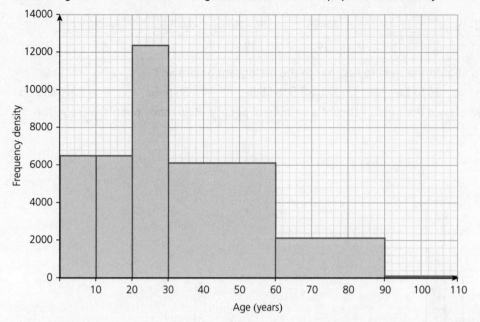

Which cumulative frequency graph shows the same data?

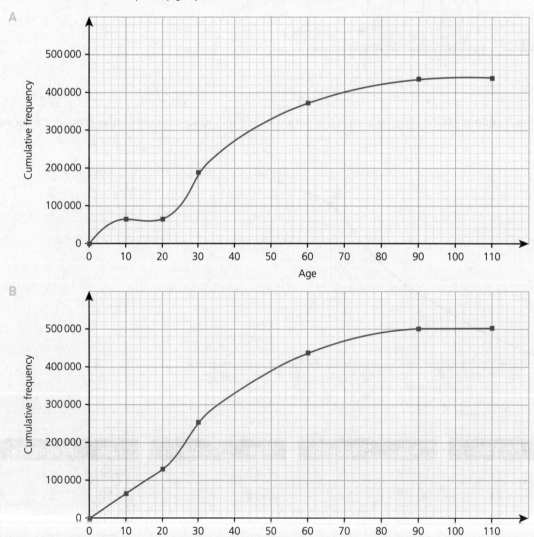

C

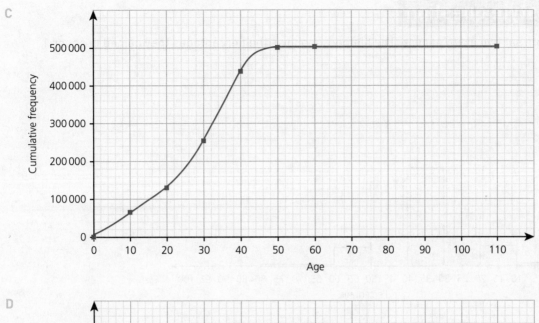

D

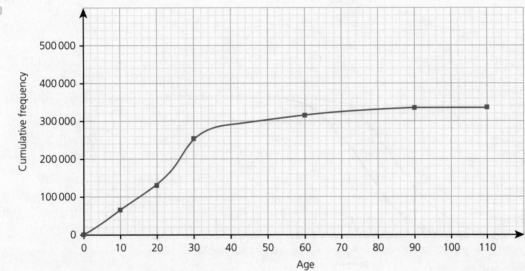

Full worked solutions online

Exam-style question

150 students sat two exam papers. The histogram shows the percentages they gained in paper 1.

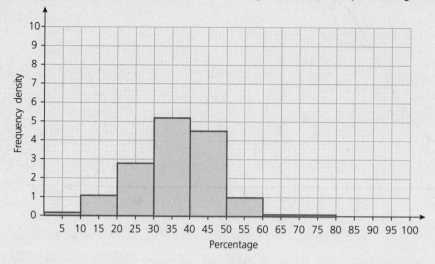

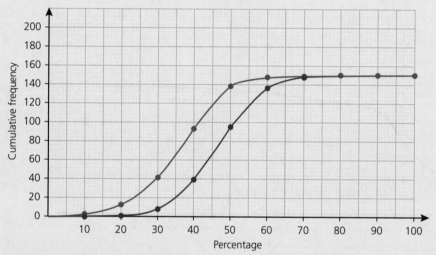

i   One of the cumulative frequency graphs is for the same data as the histogram. Determine which one.

The other cumulative frequency graph is for the data from paper 2.

ii   A   Find the number of students who got between 50% and 60% on paper 1.
  B   Find the number of students who got between 50% and 60% on paper 2.

iii Compare the students' performance on the two papers.

Short answers on page 222

Full worked solutions online

Short answers on page 222

CHECKED ANSWERS

Key facts

1 The median is the middle value of a set of data.

2 In an ordered data set of size n the median is the value of the $\frac{n+1}{2}$th data item. When n is odd this is an actual data value; for n even, the median will be the mean of two values.

3 You can find which group the median is in for a grouped frequency table in the same way.

4 To find the quartiles for a small data set, find the median first. Find the middle value of the values below the median for the lower quartile. Find the middle value of the values above the median for the upper quartile.

5 The median divides a histogram into two halves of equal area.

6 The lower quartile cuts the histogram with one quarter of the area to the left of it; the upper quartile cuts the histogram with three quarters of the area to the left of it.

7 To estimate the median from a cumulative frequency graph, read off the value with a cumulative frequency of $\frac{1}{2}n$, where n is the total number of data values.

8 The upper and lower quartiles are estimated by reading off the values with cumulative frequencies of $\frac{3}{4}n$ and $\frac{1}{4}n$, respectively.

9 In a box-and-whisker diagram, the box shows the distance between the quartiles, the whiskers show the rest.

10 The interquartile range (IQR) is a measure of spread: IQR = upper quartile − lower quartile.

11 A data value may be considered to be an outlier if it is more than $1.5 \times IQR$ above the upper quartile, or more than $1.5 \times IQR$ below the lower quartile.

Median and quartiles for a list of values

Worked example

Example 1

Find the median and quartiles of this data set: 6 4 6 7 8 11 2

You can find the median and quartiles by putting the data set into your calculator but it will be helpful if you understand how these values are calculated.

Solution

Put the data in order from smallest to largest.

Once the data are in order; this is the middle one.

The upper quartile is the middle one of the values above the median.

2 4 6 6 7 8 11

The lower quartile is the middle one of the values below the median.

Median = 6; lower quartile = 4; upper quartile = 8.

There are different recognised methods for deciding where quartiles are located; these can give slightly different answers. When you enter data into a calculator, you may find that it gives slightly different values for the lower and upper quartiles than you would get using the method above.

With an even number of data items, the median will be between two values.

The data are already in order; the middle is between 4 and 5. The median will be halfway between 4 and 5 so 4.5.

$$1\ 3\ 3\ 4\ 5\ 5\ 6\ 8$$

The lower quartile is the middle one of the four values below the median. So it is midway between the two 3s and so it is 3.

The upper quartile is the middle one of the four values above the median. Again, it lies between two numbers so will be halfway between 5 and 6.

To find the number half way between two numbers, add them and divide the answer by 2.

Median = 4.5; lower quartile = 3; upper quartile = 5.5.

Median and quartiles from a stem-and-leaf diagram

The data values in an ordered stem-and-leaf diagram are in order.

Worked example

Example 2

The stem-and-leaf diagram below shows the ages of 21 office staff. Find the median and quartiles.

Solution

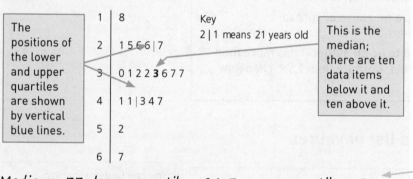

The positions of the lower and upper quartiles are shown by vertical blue lines.

1	8
2	1 5 6 6	7
3	0 1 2 2 **3** 6 7 7
4	1 1	3 4 7
5	2
6	7

Key
2 | 1 means 21 years old

This is the median; there are ten data items below it and ten above it.

Median = 33; lower quartile = 26.5; upper quartile = 42.

Common mistake: Remember the stem. The upper quartile is half way between 41 and 43.

Interquartile range (*IQR*)

- The interquartile range is a measure of spread.
  IQR = upper quartile − lower quartile.
- The interquartile range is easy to calculate and is not affected by extreme values.

Worked example

Example 3

Calculate the interquartile range for the data in Example 2.

Solution

IQR = 42 − 26.5 = 15.5

Outliers

A data item may be considered an outlier if it is more than $1.5 \times IQR$ above the upper quartile or less than $1.5 \times IQR$ below the lower quartile.

Common mistake: There is another way of seeing whether a data item is an outlier using mean and standard deviation. Use the one that is easiest to work with in any situation but don't mix them up

---

Worked example

Example 4

Could any of the ages in Example 2 be regarded as outliers?

Solution

$1.5 \times IQR = 1.5 \times 15.5 = 23.25$

$LQ - 23.25 = 26.5 - 23.25 = 3.25$

$UQ + 23.25 = 42 + 23.25 = 65.25$

Ages below 3.25 or above 65.25 can be considered to be outliers so 67 is an outlier.

You know what the quartiles are so it is easiest to use the method for outliers that uses quartiles.

Box-and-whisker diagrams

The data from Example 2 are shown in this box-and-whisker diagram.

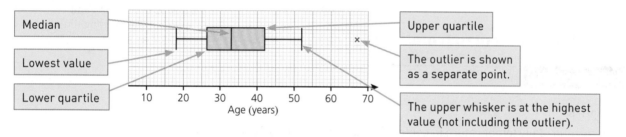

Median	Upper quartile
Lowest value	The outlier is shown as a separate point.
Lower quartile	The upper whisker is at the highest value (not including the outlier).

Age (years)

Median and quartiles from a table of data

The table below shows the number of goals scored by the winning team (or the goals scored by either team in the case of a draw) for a sample of league football matches.

Number of goals	0	1	2	3	4	5
Frequency	12	7	11	4	0	1

The median for this data set was found using a calculator in Example 3 of the section on Averages. The working for median and quartiles without a calculator is shown here to help with understanding working with grouped data.

To find the **median** from a table of data, start by adding a column for cumulative frequency to the table.

Number of goals	Frequency	Cumulative frequency
0	12	12
1	7	19
2	11	30
3	4	34
4	0	34
5	1	35

The data are the number of goals and these are in order in the table.

The cumulative frequency is 'the total frequency so far'.

35 is the total frequency for this table.

There are 35 data values in total. The median is in position number $\frac{35+1}{2} = 18$; this means it is the 18th data value. The cumulative frequency column shows you that the 18th value is one goal; so the median is one goal.

To find the median of a data set of size n, start by asking whether n is odd or even.

● If n is odd, the **median** is in position number $\frac{n+1}{2}$ once the data values are in order.

● If n is even, $\frac{n+1}{2}$ is not a whole number, this means that you will be finding the point midway between two values.

> **Hint:** For n individual data values numbered 1, 2, ..., n, the middle one is at position $\frac{n+1}{2}$.

To find the lower quartile, there are 17 data values below the median. The lower quartile is the middle one of these 17, i.e. it is in position $\frac{17+1}{2} = 9$. The lower quartile is the 9th data value so it is zero goals.

By symmetry, the upper quartile is the 9th data value after the median. The first data value after the median is number 19.

> Notice that the data value is 18 more than the number after the median. So the upper quartile is data value 9 + 18 = 27.

Data value	19	20	...	27
Number after median	1	2	...	9

The upper quartile is 2.

> **Hint:** For a grouped frequency table, you can find which groups the median and quartiles lie in using the same method as above.

Estimating the median from a histogram

Worked example

Example 5

The maximum daily temperature in central England in February 2017 is given in the histogram and table below. Find an estimate of the median.

Temperature (°C)	Frequency
$0 \leqslant T < 2$	1
$2 \leqslant T < 4$	2
$4 \leqslant T < 6$	3
$6 \leqslant T < 8$	3
$8 \leqslant T < 9$	3
$9 \leqslant T < 10$	5
$10 \leqslant T < 11$	4
$11 \leqslant T < 12$	5
$12 \leqslant T < 14$	1
$14 \leqslant T < 16$	1

> This is the same data as for Example 1 of the 'Diagrams for grouped data' section.

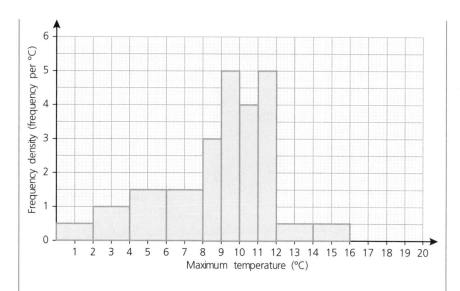

Solution

A vertical line at the median would cut the area of the histogram into two equal halves. To find an estimate of the median, start by finding the area of each bar. The frequency is the area so you can use the frequency as a measure of the area.

Half the total frequency is $28 \div 2 = 14$; this represents the area in front of the median. To get this area you need all of the first five bars and part of the sixth one.

$1 + 2 + 3 + 3 + 3 = 12$

$14 - 12 = 2$

> This is the total area of the first five bars.

> This is the area of the part of the sixth bar that is in front of the median.

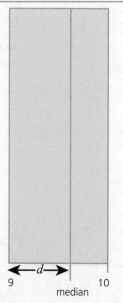

The fraction of the fourth bar that is needed is $\frac{2}{5}$ since its total frequency is 5.

The width of the bar is 1 so

$$\frac{2}{5} = \frac{d}{1}$$

$$d = \frac{2}{5} = 0.4$$

median $= 9 + d = 9 + 0.4$

median $= 9.4\,°C$

> **Hint:** Don't worry about the units when finding the areas of the bars; they will look after themselves if your scales are correct. Make sure you find the area in the same way for each bar. You can count squares or work out width times height instead of using the frequency.

> **Hint:** The fraction of width is the same as the fraction of area.

> **Common mistake:** The median is a temperature so put units on the final answer.

> **Hint:** The data have been grouped so you don't know what all the original values were. The final answer will be an estimate of the median.

Cumulative frequency curves

Worked example

Example 6

The cumulative frequency graph for the data of Example 5 is shown below. Find estimates of the median and interquartile range.

Solution

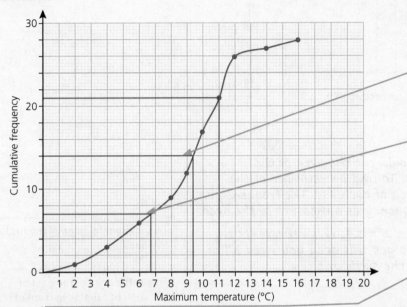

Median = 9.4 °C; LQ = 6.7 °C; UQ = 11 °C;
IQR = 11 − 6.7 = 4.3 °C.

> To estimate the median, read across from the cumulative frequency that is half the total: $(28 \div 2 = 14)$.

> To find an estimate of the LQ, read across from the cumulative frequency that is one quarter of the total: $(28 \div 4 = 7)$.

> These answers from reading from a graph are only to one decimal place but they are only estimates as you do not know the actual data values.

> **Hint:** For both the histogram and the cumulative frequency graph, the median is at the location of the middle of the continuous scale 0 to n and so it is estimated to be at $\frac{n}{2}$.

Percentiles

The quartiles (including the median) divide the data into four equal groups. Percentiles divide the data into 100 equal groups.

Worked example

Example 7

Estimate the 90th percentile of the data in Example 6.

Solution

90% of 28 is 25.2.

> First find the relevant percentage of the cumulative frequency.

> **Common mistake:** 25.2 is not the 90th percentile. It tells you the cumulative frequency to look up.

Reading from the cumulative frequency graph, the 90th percentile is 11.8 °C

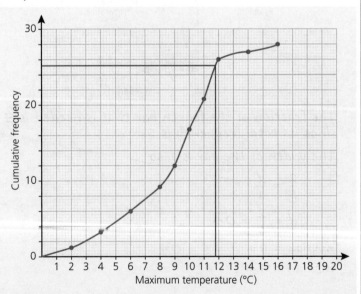

Test yourself

1 The stem-and-leaf diagram below shows the hand widths of a sample of children.

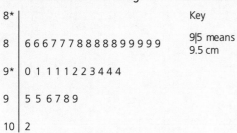

```
8*

8    6 6 6 7 7 7 8 8 8 8 8 9 9 9 9 9

9*   0 1 1 1 1 2 2 3 4 4 4

9    5 5 6 7 8 9

10   2
```

Key

9|5 means
9.5 cm

Each stem has two rows of leaves; the first row has an asterisk (*) on the stem.
Which box-and-whisker diagram correctly shows the same data?

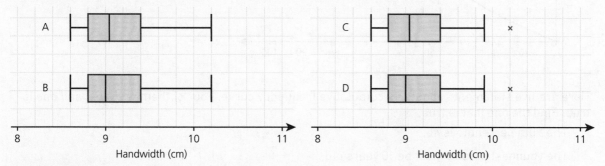

A

C ×

B

D ×

8   9   10   11   8   9   10   11

Handwidth (cm)   Handwidth (cm)

2 The waiting time, in minutes, of a sample of customers at the post office is shown in the table.

Time (t min)	$0 < t \le 1$	$1 < t \le 2$	$2 < t \le 3$	$3 < t \le 4$	$4 < t \le 5$	$5 < t \le 6$	$6 < t \le 7$	$7 < t \le 10$
Frequency	6	8	15	13	5	4	4	5

Four of the following statements are false and one is true. Which one is true?

A   There were eight customers in the sample.

B   A reasonable estimate of the median waiting time is 3.5 minutes.

C   A reasonable estimate of the total waiting time for all customers is 211 minutes.

D   The lower quartile is 15 minutes.

E   If a histogram was drawn, the final bar would be higher than the one before it.

3 Mr Brown times his bus journey from the railway station to the office. These are the times (in minutes) for ten journeys:

9   10   12   16   10   3   28   13   9   10

Four of the following statements are false and one is true. Which one is true?

A   The time of 28 minutes is an outlier and it should be ignored.

B   The mean is not the best measure of central tendency to use for this set of data.

C   28 minutes is probably a misprint for 18 minutes so the 28 should be replaced by 18.

D   3 minutes is not small enough to be an outlier, so it is a correct data value.

E   If the times had been measured and recorded accurately, there would not be any outliers.

4 In a survey, people were asked how many portions of fruit and vegetables they ate the previous day. The cumulative frequency graph on the right shows the ages of those who ate less than five portions.

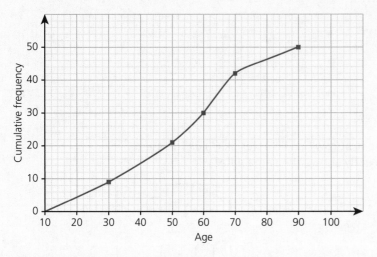

Here are five statements about this cumulative frequency curve. Four of them are false and one is true. Find the one that is true.

A   The 90th percentile is 45.

B   The youngest person must be 10 years old.

C   There were 30 people aged less than 60.

D   If you drew a histogram for the data, the second and fourth bars would be equal in height.

E   The data have been grouped so you do not know what the exact ages of the people were. This makes it impossible to tell whether there were any outliers in this case.

Full worked solutions online

CHECKED ANSWERS

Exam-style question

The cumulative percentage graph shows the annual income of households in the UK in 2015–16.

i   3.8% of households had an income over £96 000; they do not appear on the graph. What additional information is needed for these households to be added to the graph?

ii   A   Explain why you should read off from 50 on the vertical axis to estimate the median income of **all** households from the graph.

   B   Estimate the median income of **all** households from the graph.

iii   What household incomes might be considered to be outliers?

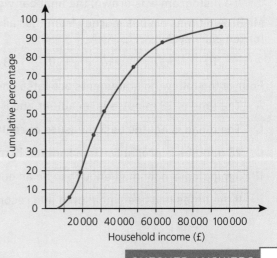

Short answers on page 222

Full worked solutions online

CHECKED ANSWERS

Standard deviation

> **Key facts**
>
> 1 The sum of squares, S_{xx}, is used in calculating several measures of spread.
>
> 2 $S_{xx} = \sum (x - \bar{x})^2 = \sum x^2 - \dfrac{\left(\sum x\right)^2}{n} = \sum x^2 - n\bar{x}^2$
>
> 3 For data in a frequency table,
>
> $$S_{xx} = \sum (x - \bar{x})^2 f = \sum x^2 f - \dfrac{\left(\sum xf\right)^2}{n} = \sum x^2 f - n\bar{x}^2.$$
>
> 4 Variance $= \dfrac{S_{xx}}{n}$.
>
> 5 Standard deviation $= \sqrt{\text{variance}} = \sqrt{\dfrac{S_{xx}}{n}}$.
>
> 6 A data item may be considered to be an outlier if it is more than two standard deviations away from the mean.

The need for a measure of dispersion

Imagine you are deciding where to go on holiday in July. You have narrowed it down to two destinations, each with an average July temperature of 27 °C. Knowing that one destination has a minimum July temperature of 21 °C and a maximum of 32 °C while the other has a minimum of 10 °C and a maximum of 41 °C would provide you with useful additional information.

The range

The range is the simplest measure of spread; it is calculated by subtracting the smallest data value from the largest data value. For the two holiday destinations above, the ranges would be:

$32 - 21 = 11$ °C and $41 - 10 = 31$ °C.

Range involves only two data points and so can be unrepresentative, particularly if there is at least one outlier and so the value of one or both of those data points is very large or very small. Other measures, such as interquartile range, are less prone to this. This section covers variance and standard deviation and the calculation of these measures is based on the whole data set and how far the data values are from the mean.

The sum of squares (of deviations from the mean)

Variance and standard deviation, considered in the next sections, use the sum of squares of deviations from the mean.

$$S_{xx} = \sum (x - \bar{x})^2 = \sum x^2 - \dfrac{\left(\sum x\right)^2}{n} = \sum x^2 - n\bar{x}^2$$

The first format of the formula makes it easier to understand what is being calculated but the second and third involve less work when doing the calculation. $(x - \bar{x})$ is the distance of a data item from the mean. Some data items will be above the mean, others will be below it; squaring $(x - \bar{x})$ makes all these values positive. The measures of dispersion that use S_{xx} are measuring how far all the data items are from the mean. If they are all near the mean, the data are not very spread out.

Worked example

Example 1

Eight students are playing a game with dice. They throw one of them and it needs to come up 6 before they can start. The following data show how many throws each of them takes to be able to start.

3   6   1   4   1   12   4   7

Calculate S_{xx}.

Solution

First find the sum of the data values.

$$\sum x = 3 + 6 + 1 + 4 + 1 + 12 + 4 + 7 = 38$$

Next find the sum of the squares of the data values.

$$\sum x^2 = 3^2 + 6^2 + 1^2 + 4^2 + 1^2 + 12^2 + 4^2 + 7^2 = 272$$

$$S_{xx} = \sum x^2 - \frac{(\sum x)^2}{n} = 272 - \frac{38^2}{8}$$

$$S_{xx} = 91.5$$

n is the number of data values (8 in this example).

Common mistake: $\sum x^2$ means 'square each data value and add them up'. Do not confuse it with $(\sum x)^2$ which would mean 'add the data values and square the answer'. These do not give the same result.

Using this form of the formula makes calculation easy; using the form with $\bar{x}$ in will mean that you might end up working with long decimals. The formulae are given at the start of the exam paper.

Hint: You could just type the data into your calculator to get values of $\sum x$, $\sum x^2$ and n. The working is only shown here so that you are reminded what your calculator is finding.

Variance and standard deviation

For the above example, variance $= \dfrac{S_{xx}}{n} = \dfrac{91.5}{8} = 11.4375\ldots$

Standard deviation $= \sqrt{\text{variance}} = \sqrt{11.4375\ldots} = 3.38$ (2 d.p.).

Hint: Your calculator will give you standard deviation automatically from a list of data or a frequency table

	1–Variable
$\bar{x}$	4.75
$\sum x$	38
$\sum x^2$	272
σx	3.38193731
sx	3.61544306
n	8 ↓

Common mistake: Look at this screen shot. There are two values shown, each of which is sometimes referred to as standard deviation: the symbols are σx and sx on this calculator. The value you need for standard deviation is σx on this calculator screen; it might have different symbols on different calculators. The value you need is calculated using $\sqrt{\dfrac{S_{xx}}{n}}$; the other one is calculated using $\sqrt{\dfrac{S_{xx}}{n-1}}$ - it is a measure of spread used in more advanced statistical work. The one you want is the smaller of the two values.

Working with summaries of data

Sometimes in examination questions you will be given a summary of the data rather than a list of all the data or a frequency table.

You need to be able to use the formulae for mean and standard deviation when given summary data.

Worked example

Example 2

70 dice are rolled and the scores are noted.

For these data, $n = 70$, $\sum x = 274$, $\sum x^2 = 1286$. Find the mean and standard deviation.

Hint: Calculators usually have more than one memory. Use the memories to keep exact values of $\bar{x}$, S_{xx}, etc., so that you can use them in later calculations.

Solution

$$\bar{x} = \frac{274}{70} = 3.914\,285.$$

$$S_{xx} = \sum x^2 - \frac{\left(\sum x\right)^2}{n} = 1286 - \frac{274^2}{70}$$

$$S_{xx} = 213.485\,714\,\dots$$

$$\text{standard deviation} = \sqrt{\frac{S_{xx}}{n}} = \sqrt{\frac{213.485\,714\,\dots}{70}}$$

$$\text{standard deviation} = \sqrt{3.049\,795\,\dots} = 1.746\,366\,\dots$$

$$\text{standard deviation} = 1.75 \ (3\ \text{s.f.})$$

Work out the mean first; remember you may need the unrounded answer for the rest of your working so put it in a calculator memory.

Work out S_{xx} before working out the standard deviation.

Hint: Do not round until you have got your final answer. You should work with the unrounded value of the mean. If you are asked to work out variance then round it sensibly but if the variance is part of your working in finding standard deviation then wait till you have worked that out before rounding, otherwise your answer will be less accurate.

Working from a frequency table

The way you enter data into a calculator to calculate standard deviation from a frequency table is the same as for mean from a frequency table; your calculator should give you both mean and standard deviation at the same time.

Worked example

Example 3

The table below shows the number of people living in a sample of 50 houses in a town. Find the mean and standard deviation.

Number of people	1	2	3	4	5
Frequency	15	17	8	7	3

		List 1	List 2	List 3	List 4
SUB					
1		1	15		
2		2	17		
3		3			
4		4			

Hint: Your calculator should allow you to type in the data as a frequency table. Make sure you know how to do this for your calculator.

Common mistake: If your calculator uses lists make sure that the frequency is set to the correct list.

1 Var	XList	: List 1
1 Var	Freq	: List 2

Solution

The mean is 2.32.

1-Variable	
$\bar{x}$	= 2.32
$\sum x$	= 116
$\sum x^2$	= 342
σx	= 1.20731106
sx	= 1.21956834
n	= 50

The standard deviation is 1.21 (3 s.f.)

Hint: Make sure you write down the correct value from the list of statistics in your calculator. You should also check that the value of n (the number of data values) is what you expect – in this case there were 50 houses; this is the total frequency.

Hint: Remember the standard deviation you want is the smaller of the two values given by your calculator.

Working from a grouped frequency table

This is the same as for an ungrouped frequency table but you need to find the midpoint of each group first.

Outliers

Outliers are unusually large or unusually small data values. They can occur for a number of reasons; they might be mistakes or they might be genuine data values that happen to be very large or small. There are several rules of thumb for identifying outliers.

A data item may be considered to be an outlier if it is more than two standard deviations away from the mean.

Worked example

Example 4

A class are trialling a test. They time how long it takes them to complete it. The mean time is 69.4 minutes and the standard deviation is 11.3 minutes. How high, or low, would a student's time have to be for it to be considered an outlier?

Solution

$\bar{x} + 2s = 69.4 + 2 \times 11.3 = 92$

$\bar{x} - 2s = 69.4 - 2 \times 11.3 = 46.8$

Times which are above 92 minutes, or below 46.8 minutes could be considered outliers.

Hint: Sometimes it is obvious that a data item is a mistake. For example, a person 2.75 m tall would be taller than the world's tallest man so such a data item would be incorrect; it should either be ignored or, if possible, corrected. On the other hand, a data item giving the height of a person as 2.21 m could be genuine but, if possible, should be checked.

Test yourself

TESTED

1 The number of goals scored by a football team in a sample of games is:

2   3   5   7   7   1   2

Find S_{xx} for this sample (some of these answers have been rounded).

A 34.53   B 36.857   C 19.5   D 141   E 2.478

2 A class of students sat a test. Their mean mark was 63. Another student sits the test later and gains a mark of 63. The mean and standard deviation are recalculated to include this mark. Four of the following statements are false and one is true. Which one is true?

  A   The mean stays the same but there is not enough information to say what happens to the standard deviation.

  B   There is not enough information to say what happens to the mean or the standard deviation.

  C   The new mean is the same as before but the standard deviation is lower.

  D   Both the mean and the standard deviation stay the same.

  E   Both the mean and the standard deviation are decreased because of the additional mark.

Questions 3 and 4 use the following data.

7 coins are tossed by each of 20 students and the number of heads for each student is noted.

$\sum x = 68, \sum x^2 = 276.$

3 Find the mean number of heads obtained on the seven coins.

  A 2.86 (3 s.f.)   B 3.4   C 4.06 (3 s.f.)   D 9.71 (3 s.f.)   E 0.294 (3 s.f.)

4 Find the variance of the number of heads on the coins.

  A 1.50 (3 s.f.)   B 1.54 (3 s.f.)   C 2.24   D 13.8 (3 s.f.)   E 14.53 (2 d.p.)

5 A sample of married women in their forties are asked how many children they have had.

Number of children	0	1	2	3	4	5	6
Frequency	13	17	38	20	8	3	1

Four of the following statements are false and one is true. Which one is true?

A $S_{xx} = 588$.

B There were 206 women in the sample.

C $n = 7$.

D $S_{xx} = 159.35$ (2 d.p.).

E The variance of children per woman was 1.636 (3 d.p.).

Full worked solutions online

CHECKED ANSWERS ☐

Exam-style question

The table below shows the heights in cm for a sample of 58 women.

Height, x cm	Frequency
$145 < x \leqslant 150$	5
$150 < x \leqslant 155$	5
$155 < x \leqslant 160$	19
$160 < x \leqslant 165$	15
$165 < x \leqslant 170$	11
$170 < x \leqslant 175$	3

i Find estimates of:
  ● the mean
  ● the standard deviation.

ii For this sample of 58 women, using the ungrouped data gives $\sum x = 9278$ and $\sum x^2 = 1\,486\,243.38$.
  Find:
  ● the mean
  ● the standard deviation.

iii Comment on any differences between your answers to parts **i** and **ii**.

Short answers on page 222

Full worked solutions online

CHECKED ANSWERS ☐

Chapter 3 Probability

About this topic

Although the basic ideas of probability can seem quite straightforward, it is important to think clearly to avoid going wrong. Probability is used in risk assessment so a good understanding can be very useful in all kinds of situations.

When using probability in real life applications, probabilities can change, depending on what has happened before – this needs care to avoid jumping to conclusions which seem intuitive but are not correct.

Before you start, remember ...

- Basic ideas of probability and calculations with fractions and decimals from GCSE.
- Substituting into formulae from GCSE.

Working with probability

REVISED

Key facts

1 The probability of an event A can often be found using
$$P(A) = \frac{\text{Number of ways } A \text{ can occur}}{\text{Total number of outcomes}}.$$
This only works if all outcomes are equally likely.

2 The expected frequency of event A in n trials is $nP(A)$.

3 $P(A') = 1 - P(A)$ where A' is the event *not A*.

4 A probability of zero means an event cannot happen; a probability of 1 means it is certain to happen.

5 Venn diagrams can be used to show either the number of outcomes or the probabilities.

6 $P(A \cup B) = P(A) + P(B) - P(A \cap B)$ where
$A \cup B$ is the union of events A or B, i.e. either A or B or both A and B occur.
$A \cap B$ is the intersection of events A and B, i.e. both A and B occur.

7 For mutually exclusive events, $P(A \cup B) = P(A) + P(B)$.

8 For independent events, $P(A \cap B) = P(A) \times P(B)$.

9 Sample space diagrams are used to display the outcomes when you have two trials happening together, each with equally likely outcomes.

	H	T
H	HH	HT
T	TH	TT

10 Tree diagrams can be used for working out the probability of two (or more) events.

11 Once you have the correct probabilities on a tree diagram, you multiply along the branches to find the probability that all the relevant events happen.

12 $P(A$ happens at least once$) = 1 - P(A$ never happens$)$.

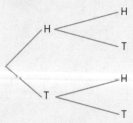

Finding probabilities using equally likely outcomes

You will almost certainly have used the following method to find probabilities:

$$P(A) = \frac{\text{Number of ways } A \text{ can occur}}{\text{Total number of outcomes}}.$$

Worked example

Example 1

What is the probability of this spinner landing on red, assuming that it is fair?

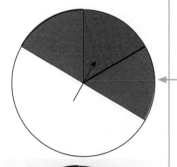

Solution

The sections are not all equal but you can put in some imaginary lines to get equal sections.

$$\text{Probability} = \frac{2}{6} = \frac{1}{3}.$$

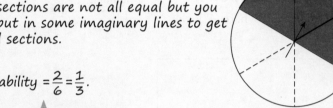

Give answers to probability questions as fractions or decimals.

Estimating probability

For the spinner from Example 1, an experiment was carried out to estimate the probability that the spinner lands on red.

The spinner was spun 1000 times and the numbers of times it came up red were used to give progressive estimates of the probability using

$$\text{Estimated probability} = \frac{\text{Number of times the spinner lands red}}{\text{Total number of trials}}.$$

The results are shown on this graph.

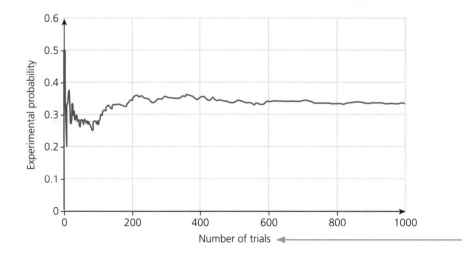

In this case, a trial consists of spinning the spinner once.

A trial is an experiment.

The graph shows the **experimental** estimates of the probability.

You have already seen in Example 1 that the **theoretical** probability of getting red is $\frac{1}{3}$.

If the experimental probability, based like this on a large number of trials, had turned out to be different from the theoretical, there would have been evidence that the spinner was **biased**.

Expected frequency of an event

In 1000 trails of spinning the fair spinner from Example 1, the expected number of times red appears is $1000 \times \frac{1}{3} = 333\frac{1}{3}$.

Impossible or certain events

- If an event cannot happen, its probability is 0.
- If an event is certain to happen, its probability is 1.

The probability that an event does not happen

Either an event happens or it does not. $P(A') = 1 - P(A)$ where A' is the event *not A*.

> **Worked example**
>
> ### Example 2
> The weather forecast estimates the probability that it will rain tomorrow as $\frac{1}{4}$. What is the probability that it will not rain?
>
> **Solution**
>
> $P(no\ rain) = 1 - \frac{1}{4} = \frac{3}{4}$

Using Venn diagrams

Venn diagrams are useful for showing the relationship between two, or more, events. They can be very helpful for finding probabilities.

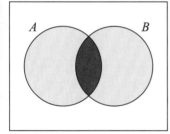
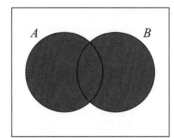

$A \cap B$: Both A and B

$A \cup B$: Either A or B or both A and B

> **Common mistake:** The expected frequency is the average number of times you would get red if you repeated the 1000 trials many times; it does not mean that you would be surprised if you did not get red $333\frac{1}{3}$ times. Since the expected frequency is an average, it need not be a whole number.

> Notice that $A \cup B$ includes the possibility that both of A and B could happen.

Venn diagrams for probability

Worked example

Example 3

The probability that it rains on a given day is 0.7, the probability that it rains and the bus is late is 0.4. The probability that it is not raining and the bus is not late is 0.1. What is the probability that the bus is late?

Solution

Let R denote the event that it rains, and B denote the event that the bus is late.

1 First fill in the probability of both events, in the middle region.

2 The probability that neither event occurs is 0.1.

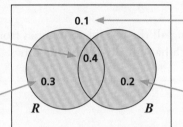

3 The probability in this region needs to add to the 0.4 in the middle to give 0.7.

4 All the probabilities need to add up to 1 so the probability for this region is 0.2.

P(bus late) = 0.4+0.2 = 0.6

The formula $P(A \cup B) = P(A) + P(B) - P(A \cap B)$ can be useful for some questions.

This formula arises because when the probabilities of event A and event B are added, the probability of the overlap is included twice.

Venn diagrams can also be used for showing the number of outcomes, or the numbers in particular sets.

Venn diagrams for numbers of objects

Worked example

Example 4

There are 21 cars in a showroom. 13 of them are silver, and 4 of the silver cars are worth over £20 000. Two of the cars are neither silver nor worth over £20 000. One of the cars in the showroom is chosen at random as a prize in a competition. What is the probability that it is worth over £20 000?

Solution

Let S denote the set of silver cars, and V denote the set of cars worth over £20 000.

1 You know the number of cars in the intersection so fill this in first.

3 There are 2 cars which are neither silver nor worth over £20 000.

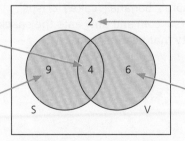

2 The total number of silver cars is 13 so there are 9 cars in this region.

4 The total number of cars is 21 so there are 6 cars in this region.

The number of cars worth over £20 000 is 4 + 6 = 10.

P(worth over £20 000) = $\frac{10}{21}$.

Mutually exclusive events

Mutually exclusive events cannot happen together.

A Venn diagram for two mutually exclusive events is shown below.

For mutually exclusive events, P(*A* or *B*) = P(*A*) + P(*B*).

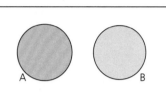

> **Common mistake:** Be careful not to add probabilities automatically when you want the probability of one event or another event; this only works when they are mutually exclusive. More generally, work with a Venn diagram or the formula
> $P(A \cup B) = P(A) + P(B) - P(A \cap B).$

Sample space diagrams

A sample space diagram is a table showing all the possible outcomes from two trials (or experiments), each of which has equally likely outcomes.

Worked example

Example 5

A fair spinner, with numbers 1 to 3 on it, is spun and a calculator chooses a random integer between 1 and 6 (inclusive). The larger number is subtracted from the smaller to give the difference. What is the most likely difference and what is the probability of getting it?

Solution

Score		Number on calculator					
		1	2	3	4	5	6
Number on spinner	1	0	1	2	3	4	5
	2	1	0	1	2	3	4
	3	2	1	0	1	2	3

> This cell in the sample space represents the event '1 on the spinner and 6 on the calculator'.

The most frequently occurring difference is 1 with a probability of $\frac{5}{18}$.

> **Hint:** If the question involves two numbers being multiplied or added, etc., then it makes your working easier to write the relevant results in the cells in the sample space.

Sample space diagrams need not involve numbers.

Worked example

Example 6

Christopher decides on his mid-morning snack by spinning a fair spinner which he has labelled with 'apple', 'banana', 'cake', 'crisps', 'chocolate', 'raisins'. He then tosses a fair coin and only eats the snack if it lands heads. What is the probability that he eats cake?

Solution

	Apple	Banana	Cake	Crisps	Chocolate	Raisins
Head (eats)						
Tail (does not eat)						

Probability of eating cake = $\frac{1}{12}$.

> Each cell in the sample space represents an equally likely outcome.

Independent events

In Example 6, the outcome on the spinner and the outcome on the coin are independent events; neither of them affects the other. The probability that Christopher eats cake can be calculated using $P(A \text{ and } B) = P(A) \times P(B)$.

P(spinner shows 'cake' and coin shows heads) = $\frac{1}{6} \times \frac{1}{2} = \frac{1}{12}$.

Tree diagrams

Tree diagrams are useful for showing the possible things that can happen and also for working out probabilities of two (or more) events.

Worked example

Example 7

A box contains three red beads and two gold beads. Apart from the colour, the beads are identical. A contestant on a game show is blindfolded. She pulls out a bead then puts it back. She then pulls out a second bead. She wins a prize if she pulls out exactly one gold bead. What is the probability that she wins a prize?

Solution

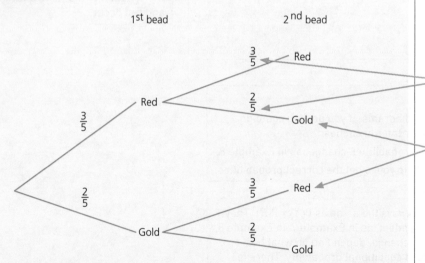

> The probabilities go on the branches.
> On each set of branches, the probabilities add to 1.

> WIN

> **Common mistake:** $P(A \text{ or } B)$ is not always the same as $P(A) + P(B)$. This only works if the events are mutually exclusive.

$P(\text{win}) = P(\text{red, gold OR gold, red}) = P(\text{red, gold}) + P(\text{gold, red})$

$P(\text{win}) = \frac{3}{5} \times \frac{2}{5} + \frac{2}{5} \times \frac{3}{5}$

$= \frac{6}{25} + \frac{6}{25} = \frac{12}{25}$

> **Hint:** Working along the branches, multiply the probabilities to find the probability that both gold and red occur.

Tree diagrams and changing probabilities

Worked example

Example 8

The rules in the game in Example 7 are changed slightly so that the first bead is not put back. What is the probability of getting exactly one gold bead now?

Solution

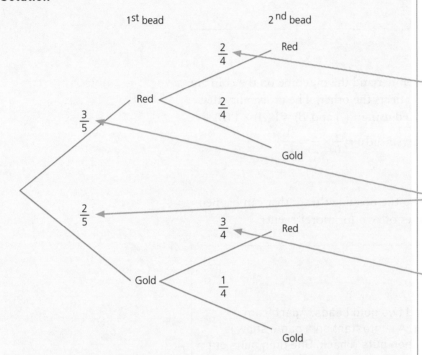

If a red bead is chosen on the first go, there are 4 beads left and 2 of them are red.

The probabilities on the first pair of branches are the same as in example 7.

If a gold bead is chosen on the first go, there are 4 beads left and 3 of them are red.

$$P(\text{win}) = P(\text{red, gold OR gold, red}) = P(\text{red, gold}) + P(\text{gold, red})$$

$$P(\text{win}) = \frac{3}{5} \times \frac{2}{4} + \frac{2}{5} \times \frac{3}{4}$$

$$= \frac{6}{20} + \frac{6}{20} = \frac{12}{20} = \frac{6}{10}$$

Working along the branches, multiply the probabilities to find the probability that both gold and red occur.

Hints:

- Do not cancel probabilities on tree diagrams; if you do, you could go wrong if the probabilities change for later branches.
- Always draw a tree diagram if the probabilities change, as in Example 8.
- Drawing a tree diagram will often help you to get the correct probabilities.

Common mistake: P(A and B) is not always the same as P(A) × P(B). They are only equal if the events are independent, as in Example 7. In Example 8, the probabilities on the branches can change, depending on what has happened before; this is an example of conditional probability. There is more about conditional probability in the next section in this chapter.

A tree diagram can be as big as you need it to be.

Worked example

Example 9

My calculator gives me a random number with 3 decimal places. Assuming that each digit is equally likely to be from 0 to 9 inclusive, what is the probability that a random number contains at least one digit 7?

Solution

There are ten equally likely possibilities for the first digit so P(7) = 0.1.

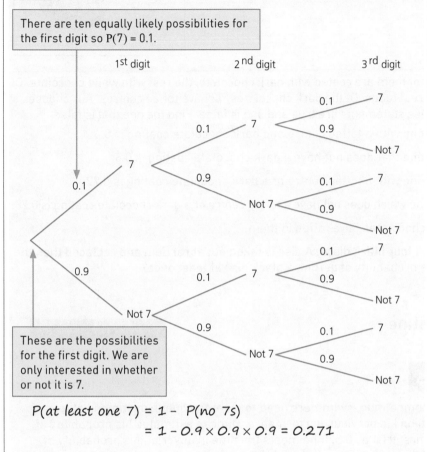

These are the possibilities for the first digit. We are only interested in whether or not it is 7.

P(at least one 7) = 1 – P(no 7s)

= 1 – 0.9 × 0.9 × 0.9 = 0.271

Hint: This is the easiest way to work out P(at least one 7).

Test yourself

TESTED

1 Rosy and Julian are taking it in turns to spin a fair spinner with the numbers 1 to 6 on it. So far it has landed 6 1 4 6 2 2 2 4.

Four of the following statements are false and one is true. Find the one that is true.

A The probability of it showing 2 on the next throw is $\frac{3}{8}$.

B It has not shown 5 yet so there is a high probability of getting 5 on the next throw.

C Getting three 2s in a row is very unlikely so there must be a typing error in the results.

D If they keep spinning the spinner, roughly one sixth of the time the spinner will show 3.

E This spinner is more likely to land showing an even number than an odd number.

2 Raffle tickets numbered 277 to 575 inclusive are put into a large container and one ticket is taken out at random. What is the probability of a number divisible by 5 being chosen?

A $\frac{1}{5}$ B $\frac{30}{149}$ C $\frac{60}{299}$ D $\frac{59}{298}$ E $\frac{59}{299}$

3 Some gardeners have phoned in to ask for their garden to appear on TV. The number of these gardens having certain features is shown in the Venn diagram below. One of these gardens will be chosen at random to be in the TV show. What is the probability that a garden with both a vegetable patch and a pond will be chosen?

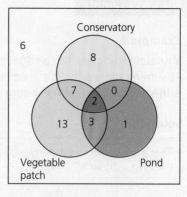

A $\frac{5}{34}$ B $\frac{13}{20}$ C $\frac{1}{8}$

D $\frac{3}{40}$ E $\frac{7}{20}$

4 A box contains 50 chocolates. 20 of them are coated with dark chocolate, the rest with white chocolate. Of these 28 are toffee-centred, the rest fudge. Of the dark chocolates, 12 have toffee centres. Alix chooses a chocolate at random. Four of these statements are true and one is false. Find the one that is false.

A   The probability of choosing one with a toffee centre and dark chocolate coating is 0.6.

B   The probability of choosing one that does not have a dark chocolate coating is 0.6.

C   The probability of choosing one with a toffee centre or a dark chocolate coating is 0.72.

D   The probability of choosing one which does not have a toffee centre or a dark chocolate coating is 0.28.

E   More than half of the white chocolates have toffee in them.

5 A jar contains six white discs and four black discs. A disc is taken out at random and replaced then the process is repeated. What is the probability of getting a black disc at least once?

A $\frac{12}{25}$ B $\frac{16}{25}$ C $\frac{2}{3}$ D $\frac{3}{4}$ E $\frac{2}{5}$

Full worked solutions online

CHECKED ANSWERS

Exam-style questions

To be in the finals of a swimming competition, swimmers need to meet the qualifying time in at least two out of three trial swims. Based on his previous performance, Duncan estimates his probability of meeting the qualifying time in the first trial as 0.8. If he meets the time in any trial, his probability of succeeding (i.e. meeting the qualifying time) in the next trial is 0.8, but if he does not meet the qualifying time in a trail, his probability of succeeding in the next trial is reduced to 0.6.

i Find the probability that Duncan will succeed in meeting the qualifying time in at least one trial.

ii Find the probability that Duncan will be in the finals of the competition.

Short answers on page 222

Full worked solutions online

CHECKED ANSWERS

Conditional probability

REVISED

Key facts

1   $P(A|B)$ is read "the probability of A given B"; it means the probability that A occurs given that B has already occurred.

2   $P(A|B) = \frac{P(A \cap B)}{P(B)}$ where $A \cap B$ is the event A and B.

3   $P(B|A) = P(B) \Leftrightarrow B$ and A are independent and, in that case, $P(A \cap B) = P(A)\,P(B)$.

What is conditional probability?

Sometimes having additional information changes the probability. This can be most easily seen in medical applications, for example older people are more at risk of some conditions than the general population.

> **Worked example**
>
> ### Example 1
>
> 100 students were surveyed to find out whether they have a part-time job. The results are shown in the table below. A student is chosen at random from this group.
>
> | | Female | Male | Total |
> |---|---|---|---|
> | Part-time job | 54 | 35 | 89 |
> | No part-time job | 4 | 7 | 11 |
> | Total | 58 | 42 | 100 |
>
> J stands for the event 'the student has a part-time job'.
>
> F stands for the event 'the student is female'. Find the following probabilities.
>
> **i** $P(J)$ **ii** $P(F)$ **iii** $P(J \cap F)$ **iv** $P(J|F)$

This is a conditional probability, the probability that the student has a part-time job given that the student is female.

> **Solution**
>
> **i** 89 out of the 100 students have a part-time job.
>
> $P(J) = \frac{89}{100} = 0.89$.
>
> **ii** 58 out of the 100 students are female.
>
> $P(F) = \frac{58}{100} = 0.58$.
>
> **iii** 54 out of the 100 students have a part-time job and are female. $P(J \cap F) = \frac{54}{100} = 0.54$.
>
> **iv** You are given that the student is female. 54 out of the 58 female students have part-time jobs.
>
> $P(J|F) = \frac{54}{58} \approx 0.931$.

Hint: Using the formula $P(A|B) = \frac{P(A \cap B)}{P(B)}$ for Example 1(iv) would give $P(J|F) = \frac{P(J \cap F)}{P(F)}$, so you could get the answer to (iv) from the answers to (iii) and (ii) $P(J|F) = \frac{0.54}{0.58} \approx 0.931$. This is the same working as in the solution to the left but using probabilities instead of frequencies.

Some questions are about general events where you have to work with the formulae.

> **Worked example**
>
> ### Example 2
>
> $P(A) = 0.6$, $P(B) = 0.5$, $P(A|B) = 0.8$. Find $P(A \cap B)$.
>
> **Solution**
>
> You can find $P(A \cap B)$ using the conditional probability formula.
>
> $P(A|B) = \frac{P(A \cap B)}{P(B)}$
>
> $0.8 = \frac{P(A \cap B)}{0.5}$
>
> $P(A \cap B) = 0.8 \times 0.5 = 0.4$

Put the probabilities you know into the formula.

Hint: Notice that the formula for $P(A|B) = \frac{P(A \cap B)}{P(B)}$ has $P(B)$ in the denominator.

Conditional probability and Venn diagrams

> **Worked example**
>
> ### Example 3
> P(A) = 0.6, P(B) = 0.5, P(A ∩ B) = 0.4. Work out P(A ∪ B) and P(B|A).
>
> #### Solution
>
> You can put the probabilities into a Venn diagram.
>
> $P(A \cup B) = 0.2 + 0.4 + 0.1 = 0.7$
>
>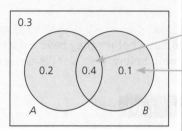

Hint: Start with P(A ∩ B) in the intersection then work outwards.

Hint: P(B) = 0.5 so the probability for this section is 0.5 − 0.4 = 0.1.

> **Hints:**
> - You could use the formula $P(A \cup B) = P(A) + P(B) - P(A \cap B)$ without drawing a Venn diagram.
> - P(A|B) does not appear in the Venn diagram but it could be worked out using $P(A|B) = \dfrac{P(A \cap B)}{P(B)} = \dfrac{0.4}{0.5}$.
> - Notice that $P(B|A) = \dfrac{P(A \cap B)}{P(A)} = \dfrac{0.4}{0.6} = \dfrac{2}{3}$ and this is different to P(A|B).

Conditional probability and tree diagrams

The idea of putting conditional probabilities on the branches of tree diagrams has been covered in the previous section (see Example 8).

Deciding whether events are independent

> **Worked example**
>
> ### Example 4
> A random sample of 200 plants of a particular variety of flower were examined. For these plants, A is the event 'plant has dark leaves' and B is the event 'plant has red flowers'. The Venn diagram shows the numbers of plants with these characteristics. Find:
>
> i   P(A)
> ii P(B)
> iii P(A ∩ B).
> iv State whether events A and B are independent, giving reasons for your decision.

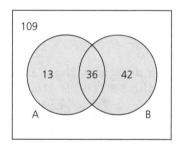

Common mistake: Be careful not to assume that P(A ∩ B) is P(A) × P(B). You can work out P(A ∩ B) directly from the Venn diagram in Example 4.

> #### Solution
>
> i $P(A) = \dfrac{13 + 36}{200} = 0.245$
>
> ii $P(B) = \dfrac{36 + 42}{200} = 0.39$
>
> iii $P(A \cap B) = \dfrac{36}{200} = 0.18$
>
> iv If the events are independent then
> $P(A \cap B) = P(A) \times P(B)$.
> $P(A) \times P(B) = 0.245 \times 0.39 = 0.09555$ but
> $P(A \cap B) = 0.18$.
>
>   These are not equal so the events are not independent.

Hint: If you are asked to explain whether two events are independent you are not expected to write a paragraph about why they should not affect the probability of each other happening. You should use one of the facts that, for independent events, P(B|A) = P(B) or P(A ∩ B) = P(A) P(B), explaining briefly what you are doing, as in the solution on the left.

Test yourself

1 A jar contains 8 red discs and 5 green discs. Two discs are picked at random. The first one is not replaced. What is the probability that the second disc is red, given that the first disc is green?

A $\dfrac{1}{2}$ B $\dfrac{2}{3}$ C $\dfrac{7}{12}$ D $\dfrac{8}{13}$

E It is impossible to work out without more information.

2 A researcher is conducting a survey by asking questions in the street. She keeps a record of the number of people who were willing to answer her questions and of those who were not. The results are shown in the table below. Assuming that these results are typical, what is the probability that someone is willing to answer given that the person is female?

	Male	Female
Willing to answer	51	81
Not willing to answer	70	61

A $\dfrac{81}{263}$ B $\dfrac{132}{263}$ C $\dfrac{142}{263}$ D $\dfrac{81}{132}$ E $\dfrac{81}{142}$

3 A fair coin is tossed. If it lands heads, a ball is chosen at random from jar X. If it lands tails a ball is chosen at random from jar Y. There are 6 green balls and 5 blue balls in jar X. There are 2 green balls and 8 blue balls in jar Y. Four of the following statements are true and one is false. Find the one that is false.

A The probability of choosing a green ball is $\dfrac{41}{110}$.

B The events 'the coin lands tails' and 'a green ball is chosen' are independent because the balls cannot know how the coin has landed.

C The probability of the coin landing heads and then choosing a blue ball is $\dfrac{5}{22}$.

D The probability of choosing a green ball given that the coin lands heads is $\dfrac{6}{11}$.

E A blue ball is more likely to be chosen than a green ball.

4 You are given that $P(A) = 0.5$, $P(B) = 0.4$ and $P(B|A) = 0.3$. What is $P(A \cup B)$?

A 0.15   B 0.2   C 0.7   D 0.75   E 0.9

5 Munchy Biscuits are sold in cartons. During a promotion, the lids of the cartons are coloured and numbered on the inside. Customers win if the lid is red with number 10. 30% of the lids are red. 20% of the lids have a number 10. For the red lids, 10% of them are numbered 10. What is the probability of winning?

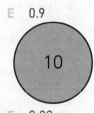

A 0.5   B 0.1   C 0.06   D 0.03   E 0.02

Full worked solutions online

Exam-style question

Bill's cat has eight kittens: three female and five male. Alex is going to have two of them.

Two of them are selected at random for Alex.
i What is the probability that two females are selected?
ii What is the probability that one of each gender is selected?
iii What is the probability that two females are selected given that both kittens chosen are of the same gender?

Short answers on page 222

Full worked solutions online

Chapter 4 The binomial distribution

About this topic

Many of the situations where you can use tree diagrams involve two possibilities at each stage. If the probabilities stay constant throughout the tree diagram, this is a binomial situation. The binomial probability model will be developed in this chapter to allow you to calculate such probabilities quickly.

Using the binomial probability formula enables you to work out the probability of one outcome. If you want several outcomes combined together, it is much quicker to use the probability functions on your calculator.

Before you start, remember ...

- Probability; tree diagrams; (revised in chapter 3).

Introducing the binomial distribution

Key facts

1 In a binomial situation, these criteria apply:
- You are conducting an experiment or trial n times (a fixed number), e.g. tossing a coin eight times.
- There are two outcomes, which you can think of as 'success' and 'failure', e.g. heads and tails.
- The probability of 'success' is the same each time (symbol p). The probability of 'failure' is $q = 1 - p$.
- The probability of 'success' on any trial is independent of what has happened in previous trials.
- The random variable, X, is 'the number of successes'.

2 You write $X \sim B(n, p)$ to show that X has a binomial distribution. The values of n and p are called the **parameters** of the binomial distribution.

3 $P(X = r) = {}_nC_r q^{n-r} p^r$
where $q = 1 - p$
$r = 0, 1, 2, ..., n$

$${}_nC_r = \frac{n!}{r!(n-r)!}$$

Worked example

Example 1

A biased coin has probability 0.4 of landing heads. It is tossed three times. What is the probability that it lands heads twice?

Solution

This can be solved by using a tree diagram.

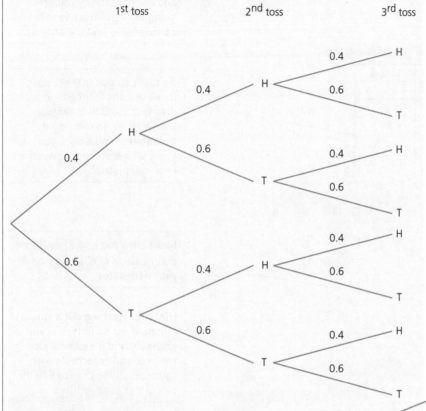

P(2 heads) = P(HHT or HTH or THH)

P(2 heads) = P(HHT) + P(HTH) + P(THH)

P(HHT) = 0.4 × 0.4 × 0.6

P(HTH) = 0.4 × 0.6 × 0.4

P(THH) = 0.6 × 0.4 × 0.4

P(2 heads) = 3 × 0.096 = 0.288

Hint: Each of these three probabilities is $0.6 \times 0.4^2 = 0.096$.

Hint: Notice that the probability is $3 \times 0.6 \times 0.4^2$. There are three routes through the tree diagram with one tail and two heads. The tail has probability 0.6 and each head has probability 0.4.

For the random variable X = number of heads, using the binomial probability formula with $n = 3$ and $p = 0.4$ gives this probability in one calculation.

$$P(X = r) = {}_n C_r q^{n-r} p^r \quad \text{where} \quad q = 1 - p$$
$$P(X = 2) = {}_3 C_2 \times 0.6^{3-2} \times 0.4^2$$
$$= 3 \times 0.6 \times 0.4^2$$
$$P(X = 2) = 0.288$$

This part of the formula tells you how many relevant routes there would be through a tree diagram.

This part of the formula tells you what the probability for each relevant route would be if you used a tree diagram.

Hint: When using the binomial probability formula, start by writing down what X, n, p and q stand for. Write down the formula and substitute the correct numbers into it then use your calculator to find the answer.

Using a calculator

It is possible to get binomial probabilities from calculators by inputting values of n, p and r. If you use a calculator, always write down the binomial distribution you are using and the probability you are calculating.

For Example 1, you should write the following:

X = number of heads

$X \sim B(3, 0.4)$

$P(X = 2)$

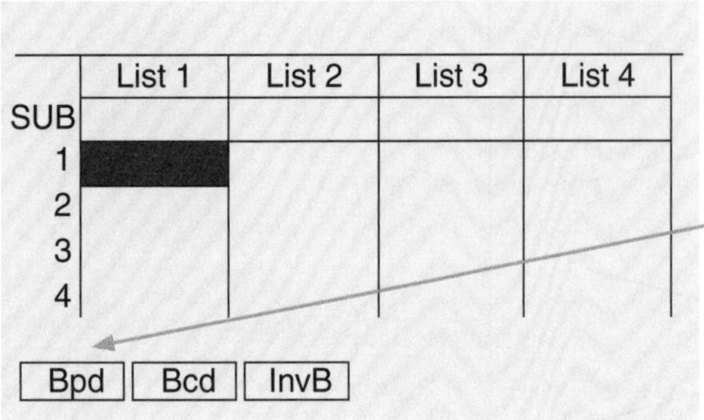

	List 1	List 2	List 3	List 4
SUB				
1				
2				
3				
4				

Bpd	Bcd	InvB

Binomial	P.D
Data	Variable
X	2
Numtrial	3
P	0.4
Save Res	None
Execute	

$P(X = 2) = 0.288$

Using the binomial probability formula

The binomial probability formula speeds up some probability calculations and allows you to deal with questions where it would take too much time and space to draw a tree diagram.

Worked example

Example 2

Adam throws ten dice and gets a score of 6 seven times. He wonders if there is something wrong with the dice. What is the probability of getting seven 6s when throwing ten fair dice?

Solution

X = number of 6s

$n = 10$, $p = \frac{1}{6}$

You are interested in the number of 6s.

10 dice are thrown; n is the number of trials.

If the dice are fair, the probability of a 6 each time is $\frac{1}{6}$.

$$q = 1 - \frac{1}{6} = \frac{5}{6}$$

This is the probability of not getting a 6.

$$P(X = r) = {}_nC_r\, q^{n-r} p^r$$

$$P(X = 7) = {}_{10}C_7 \times \left(\frac{5}{6}\right)^3 \times \left(\frac{1}{6}\right)^7 = 0.000\,248 \text{ (3 s.f.)}$$

You want the probability of seven 6s so $r = 7$.

Finding the probability of 'at least one'

Worked example

Example 3

A multiple choice test has ten questions, each with five possible answers. A student guesses all the answers. What is the probability he gets at least one correct?

Hint: If you are asked to find the probability of at least one 'success', it is usually quickest to find the probability of no successes and subtract this from 1. There are either no successes or there is at least one, so these probabilities add to 1.

Solution

X = number of correct answers

$n = 10, p = \frac{1}{5}, q = \frac{4}{5}$

$P(X = r) = {}_nC_r\, q^{n-r} p^r$

$P(X = 0) = {}_{10}C_0 \left(\frac{4}{5}\right)^{10} \left(\frac{1}{5}\right)^0 = 0.107\,374 \ldots$

$P(\text{at least one right}) = 1 - 0.107\,374 \ldots = 0.892\,625 \ldots$
$$= 0.893 \text{ (3 s.f.)}$$

Hint: Find the probability of getting none right first.

Common mistake: It is much easier to read an answer given to about 3 significant figures. Even though the probability with more decimal places is more accurate, it is helpful to round it as well as giving the full answer.

Recognising situations which are binomial

In an exam paper of mixed questions, it is not always obvious when you should use the binomial probability formula. It is even less obvious when using probability in real life. The key facts at the start of this section will help you to decide if it is a binomial situation.

Worked example

Example 4

5% of the population in an area are vegetarian. 30 people from the area will come to a meeting where lunch will be served. What is the distribution of the number of vegetarians at such a meeting? What assumptions have you made in answering this question?

n is the number of trials. There are 30 people coming to the meeting and checking whether each one is vegetarian is a 'trial'.

Solution
X = number of vegetarians at the meeting
$n = 30$
$p = 0.05$
$X \sim B(30, 0.05)$

The proportion of vegetarians in the population is 0.05.

This is how you write that X has a binomial distribution with $n = 30$ and $p = 0.05$.

Assumptions: The probability of each person being vegetarian is 0.05. This probability is independent of whether anyone else at the meeting is vegetarian or not.

This may not be the case; for example, the meeting might be about a subject which is of greater interest to vegetarians.

Worked example

Example 5

Half the students in a class of 30 are girls. The teacher chooses students at random to answer questions but does not ask the same student twice in a lesson. He asks 12 questions in a lesson. Explain why you could not use a binomial distribution to find the probability that six girls and six boys are questioned.

Common mistake: Watch out very carefully to see if the probability of 'success' is the same each time, and independent of what has happened in previous trials.

Solution

The probability of the first student being a girl is $\frac{15}{30} = 0.5$

but, once the first student has been asked, the probability of the second student being a girl will be different because there are now only 29 students to choose from. If the first student is a girl, there are only 14 girls left. For a binomial distribution, the probability must be the same each time.

Common mistake: Notice that in the situation in Example 5, the students are being selected from 30 students so the probabilities change, but in Example 4, people are being selected from a much larger population so the probabilities are assumed to be equal – they would change a bit but the binomial model is still a good approximation.

Test yourself

TESTED

1. A radio station is having a phone-in programme. They guarantee that exactly 70% of calls, chosen at random, will be answered. Five friends ring up. What is the probability that exactly two of them have their calls answered?

   A 0   B 0.01323   C 0.1323   D 0.3087   E 0.49

2. 'One-fifth of all chocolate bars contain a winning ticket' says the advert. Mark is determined to get a winning ticket. He buys ten chocolate bars. What is the probability that he has one or two winning tickets? (You should assume that the advert is true and that each bar of chocolate is equally likely to contain a winning ticket. Some answers are rounded to 4 d.p.).

   A 0.0336   B 0.0811   C 0.24   D 0.5704   E 1

3. There are two tills in a shop. One of the assistants thinks that each customer is equally likely to go to either till. If she is right, what is the probability that, for the next ten customers, five go to each till?

   A $\frac{1}{1024}$ B $\frac{63}{256}$ C $\frac{1}{32}$ D $\frac{1}{16}$ E 1

4. 95% of the ball point pens produced in a factory work. They sell them in packs of six pens and give a money back guarantee if a customer buys a pack that contains any pens that do not work. What percentage of packs will they have to refund money on? (You should assume that the faulty pens are thoroughly mixed with the working pens before packing.)

   A 30%   B 26.5%   C 99.9%   D 23.2%   E 5%

5 The binomial distribution cannot be used for four of the random variables described below. It can be used for one of them. Find the one where the binomial distribution can be used.

   A   Five fair dice are thrown. The random variable is the total score on the dice.

   B   A bag contains a fixed number of balls. Some are black and the rest are white. A ball is taken without looking and its colour noted. It is replaced and the balls are mixed. This process takes place five times. The random variable is the number of times a white ball is seen.

   C   A fair coin is tossed until it lands heads. The random variable is the number of tosses.

   D   Students take a 40-question multiple choice test, for which they have revised. Each question has 5 possible answers. The random variable is the number of questions a student gets right.

   E   Rose opens a box of chocolates and eats them at random. There are 14 chocolates in the box, 4 are plain and the rest are milk. Rose eats 8 chocolates. The random variable is the number of plain chocolates she eats.

Full worked solutions online

CHECKED ANSWERS ☐

Exam-style question

10% of the population are dyslexic. A school has classes which contain 30 students.
i   Use a binomial probability model to find the following probabilities.
   A   The probability that a class has one dyslexic student in it.
   B   The probability that a class has at least one dyslexic student in it.
ii State one assumption needed for the binomial model you used to be valid.

Short answers on page 223

Full worked solutions online

CHECKED ANSWERS ☐

Cumulative probabilities and expectation

REVISED ☐

Key facts

1   The expectation of a binomial random variable is $E(X) = np$. The variance is npq.

2   Make sure you know whether your calculator works out $P(X \leqslant x) = P(X = 0) + P(X = 1) + ... + P(X = x)$ or $P(a \leqslant X \leqslant b)$ and that you can use it to answer any question which involves the probability of a range of values of a binomial random variable.

The expectation of a binomial random variable

For a binomial discrete random variable, the expected value, or mean, is given by the following formula.

For $X \sim B(n, p)$, $E(X) = np$.

Worked example

Example 1

A bank claims that 90% of callers to its helpline wait less than five minutes to speak to an adviser. A random sample of 50 callers is surveyed to find out how long it took before they were able to speak to an adviser. What is the expected number in the sample that waited at least five minutes?

Solution

For a random sample, it is reasonable to assume that the probability for each caller of waiting longer than five minutes is the same and independent of other callers.

X = number of callers in the sample who waited at least five minutes

$X \sim B(50, 0.1)$

$E(X) = np = 50 \times 0.1 = 5.$

> Remember to say what any letters you introduce stand for.

> 90% less than five minutes means 10% at least five minutes.

> Show that you know this is a binomial distribution and give the parameters.

Common mistakes:

- If all callers one evening were surveyed instead of a random sample, it would not be reasonable to assume that the probability of each one waiting at least five minutes was the same.

  If one caller takes a long time on the phone, this will make the waiting time of the ones who come after him (or her) longer.

- The expected value might not have a very high probability of occurring and it need not be a whole number.

Worked example

Example 2

For the sample in Example 1, what is the probability that exactly two callers in this sample waited at least 5 minutes?

Solution

$X \sim B(50, 0.1)$

$P(X = r) = {}_nC_r \, p^r q^{n-r}$ where $q = 0.9$

$P(X = 2) = {}_{50}C_2 \times 0.1^2 \times 0.9^{48} = 0.077942...$
$= 0.078$ (to 3 s.f.)

> Example 1 shows what the letters stand for, this is not repeated here.

> Putting the numbers into the binomial probability formula then using a calculator to get the answer.

Using your calculator to work out the total probability of range of values for a binomial distribution

The binomial probability formula is quick to use if you just want a single probability. To save work when dealing with situations where several probabilities would need to be added, you should use the probability functions on your calculator. This is shown in the next two examples.

> **Hint:** Some calculators will work out $P(X \leqslant x) = P(X = 0) + P(X = 1) + ... + P(X = x)$ for a binomial distribution; these are usually scientific calculators. Other calculators will work out $P(a \leqslant X \leqslant b)$ for any range of values for a binomial distribution; these are usually graphical calculators. Make sure you know what your calculator does and can use it to answer any question which involves the probability of a range of values of a binomial random variable.

Worked example

Example 3

An office has eight identical printers. On any given day, the probability of any one of them being faulty is 0.15, independently of the others. What is the probability that, on a given day, three or fewer of the printers are faulty?

Solution

X = the number of faulty printers on a given day.

$n = 8$, $p = 0.15$ where p is the probability of a printer being faulty.

X has a binomial distribution as the probability of each printer being faulty is the same so $X \sim B(8, 0.15)$.

You want $P(X \leq 3)$.

> **Hint:** Write what you are asked to find in symbols. Now use your calculator to get the answer.

Using $P(X \leq x)$	Using $P(a \leq X \leq b)$

Using $P(X \leq x)$

```
STAT–REG DISTR
4      : Binomialpdf
5      : Binomialcdf
6      : Poissonpdf
```

To work out the total probability for a range of values for a binomial distribution, you need the Binomialcdf option.

> **Hint:** The c in cdf stands for cumulative – it is the same idea as for cumulative frequency but adding up probabilities instead of frequencies.

```
Binomialcdf              ↑
TRIALS=n          = 8
P(SUCCESS)        = 0.15
X                 = 3
```

Entering 3 for x will give $P(X \leq 3)$.

Using $P(a \leq X \leq b)$

	List 1	List 2	List 3	List 4
SUB				
1				
2				
3				
4				

```
Bpd    Bcd    InvB
```

To work out the total probability for a range of values for a binomial distribution, you need the Bcd option.

> **Hint:** The c in Bcd stands for cumulative – it is the same idea as for cumulative frequency but adding up probabilities instead of frequencies.

```
Binomial        C.D
Data      :  Variable
Lower     :  0
Upper     :  3
Numtrial  :  8
P         :  0.15
Save Res  :  None
```

$P(X \leq 3) = P(0 \leq X \leq 3)$ so the lower value is 0 and the upper value is 3.

Using P($X \leqslant x$)	Using P($a \leqslant X \leqslant b$)
Binomialcdf VALUE = 0.97864753 STORE : yztabcd SOLVE AGAIN QUIT	Binomial C.D P= 0.97864753
Now write down the answer, rounding sensibly – remember the probability of 0.15 given in the question will be an estimate. P($X \leqslant 3$) = 0.9786	Now write down the answer, rounding sensibly – remember the probability of 0.15 given in the question will be an estimate.

Phrase	Less than 3	No more than 3	Up to 3	More than 3	No less than 3	At least 3
Symbols	$X < 3$	$X \leqslant 3$	$X \leqslant 3$	$X > 3$	$X \geqslant 3$	$X \geqslant 3$
Values (X is an integer)	0, 1, 2	0, 1, 2, 3	0, 1, 2, 3	4, 5, ...	3, 4, ...	3, 4, ...

Common mistake: Be careful when writing the probability you want as an inequality. Some commonly used phrases, and how they translate to symbols, are listed in the table below.

Hint: You may find it helpful to write down the possible values that you are trying to find the probability of in order to help you to see how to enter the information into your calculator.

Worked example

Example 4

18 fair dice are thrown. What is the probability that there are more than six sixes?

Solution

X = the number of 6s; this has a binomial distribution as the dice are all fair.

$n = 18$, $p = \frac{1}{6}$, where p = probability of throwing a 6.

$X \sim B\left(18, \frac{1}{6}\right)$.

You want P($X > 6$). Possible values are 7, 8, 9, 10, ..., 18. Hence, not 0, 1, 2 ..., 6,

Hint: Write down the possible values that you are trying to find the probability of so you can see how to enter the information into your calculator.

Using P($X \leqslant x$)	Using P($a \leqslant x \leqslant b$)

Binomialcdf
TRIALS   = n=18
P(SUCCESS) = 0.16666..
X   = 6
                    CALC

Enter $\frac{1}{6}$ for P(success)
Entering 6 for x will give P($X \leqslant 6$)

Binomial   C.D
Data   : Variable
Lower   : 7
Upper   : 18
Numtrial   : 18
P   : 1÷6
Save Res   : None

Enter $\frac{1}{6}$ for p
The lower value is 7 and the upper value is 18

Binomialcdf
VALUE = 0.979361064
STORE : yztabcd
SOLVE AGAIN QUIT

P($X > 6$) = 1 – P($X \leqslant 6$)
          = 1 – 0.9794 = 0.0206

Binomial   C.D

    P = 0.02063893

P($X > 6$) = 0.0206

Worked example

Example 5

For the random variable, X, in Example 4, find P($3 \leqslant X < 7$).

Solution

P($3 \leqslant X < 7$)

Hint: The possible values are 3, 4, 5, 6.

Using P($X \leqslant x$)	Using P($a \leqslant X \leqslant b$)

Binomialcdf
TRIALS   = n=18
P(SUCCESS) = 0.16666..
X   = 6
                    CALC

Enter $\frac{1}{6}$ for P(success)
Entering 6 for x will give P($X \leqslant 6$)

Binomial   C.D
Data   : Variable
Lower   : 3
Upper   : 6
Numtrial   : 18
P   : 0.16666666

The lower value is 3 and the upper value is 6

Remember that lower and upper values are inclusive for this kind of calculator

```
Binomialcdf
VALUE = 0.979361064
STORE : yztabcd
SOLVE AGAIN QUIT
```

Either write down this value or store it in a calculator memory

```
Binomialcdf
TRIALS       = n =18
P(SUCCESS) = 0.16666..
X            = 2
                  CALC
```

```
Binomial   C.D
    P = 0.57670675
```

$P(3 \leqslant X < 7) = 0.5767.$

```
Binomialcdf
VALUE    = 0.402654317
STORE    : yztabcd
SOLVE AGAIN QUIT
```

This gives $P(X \leqslant 2) = 0.40265...$

$P(3 \leqslant X < 7)$
$= P(X \leqslant 6) - P(X \leqslant 2)$
$= 0.9794 - 0.4027$
$= 0.5767$

Common mistake: Remember to write down what probability you are calculating at each stage before getting the answer on the calculator so that you don't lose track of what you have got the answer to.

Test yourself

1 In the game 'find the lady' the player has to say which one of three shuffled cards is the queen. The cards are face down. A player plays this game eight times. What is the expected number of times he wins, assuming that he is guessing?

A 2   B 2.5   C $2\frac{2}{3}$ D 3   E 4

2 X is a binomial random variable. Below are five pairs of inequalities; one in words and one in symbols. Four of the pairs are equivalent and one pair is not equivalent. Find the pair that is not equivalent.

A X is no more than 6; $X \leqslant 6$.

B X is at least 5; $X \geqslant 5$.

C X is more than 6; $X > 6$.

D X is not less than 7; $X \geqslant 7$.

E X is at most 7; $X > 7$.

3 X is a binomial random variable. Below are five statements about probabilities involving inequalities. Four of them are true and one is false. Find the one that is false.

A $P(X \geqslant 4) = 1 - P(X \leqslant 4)$

B $P(X < 3) = P(X \leqslant 2)$

C $P(X > 5) = 1 - P(X \leqslant 5)$

D $P(3 < X < 8) = P(X \leqslant 7) - P(X \leqslant 3)$

E $P(2 \leqslant X \leqslant 6) = P(X \leqslant 6) - P(X \leqslant 1)$

4 In the UK, 20% of children aged 2–15 years have asthma. Two adults are planning to take 12 children from this age group on a trip. Assuming that the children are a random sample from the population, what is the probability that no more than 2 out of the 12 children will suffer from asthma?

A   0.4896   B   0.2835   C   0.4417   D   0.5583   E   0.5584

5 There are ten hurdles in a race. A particular runner has a 65% chance of clearing any one of them. This probability is not affected by whether or not she has cleared the previous hurdles. What is the probability that she clears more than half the hurdles, but not all of them? (All answers are rounded to 3 d.p.)

A   0.500   B   0.892   C   0.738   D   0.741   E   0.905

Full worked solutions online

CHECKED ANSWERS

Exam-style question

23% of people who live in Wales can speak Welsh.

A researcher will interview a random sample of 50 people who work for a particular employer.

i   Give one reason why a binomial distribution with $n = 50$, $p = 0.23$ might not be a suitable model for the number of people in the sample who speak Welsh.

For the rest of this question you should assume that a binomial distribution with $n = 50$, $p = 0.23$ is a suitable model for the number of people in the sample who speak Welsh.

ii Find the probability that at least five people in the sample speak Welsh.

iii The researcher interviews the people one at a time. What is the probability that the first one he interviews who speaks Welsh is the 5th person?

Short answers on page 223

Full worked solutions online

CHECKED ANSWERS

Chapter 5 Statistical hypothesis testing using the binomial distribution

About this topic

A common and widely used statistical method is to take a sample and to use it to draw possible conclusions (inferences) about the population from which it is drawn. However, there is always a possibility that those conclusions, based on sample data, are wrong or inaccurate. A hypothesis test uses probability to assess whether a proposed model is consistent with the sample data. To take a simple example, if a coin lands heads 14 times and tails 6 times, is this consistent with the coin being fair or is there enough evidence to convince us that it is biased?

In a 1-tail test, you are looking for evidence in a particular direction, for example that the coin is biased towards heads. For a 2-tail test you are looking for evidence of any difference rather than for evidence of a difference in a particular direction.

Before you start, remember ...

- Binomial probability, including use of calculators (this is revised in Chapter 4).

Introducing hypothesis testing using the binomial distribution (1-tail tests)

> **Key facts**
>
> 1. In hypothesis testing for a binomial distribution, you use evidence from a sample to make a decision about the probability of 'success' for the whole population.
> 2. For a hypothesis test using the binomial distribution, the null hypothesis is written in the form $H_0 : p = 0.3$ (or some other specific value)
> 3. There should be a statement defining what p stands for, e.g. p = the probability of a light bulb lasting less than 100 hours.
> 4. For a 1-tail test, the alternative hypothesis that goes with the example null hypothesis above is EITHER $H_1 : p > 0.3$ OR $H_1 : p < 0.3$. 1-tail tests are covered in this section.
> 5. For a 2-tail test, the alternative hypothesis that goes with the example null hypothesis above is $H_1 : p \neq 0.3$. 2-tail tests are covered in the next section.
> 6. You can never be certain of the outcome from a hypothesis test but you can say that it is very unlikely that the null hypothesis is true. The **significance level** is the probability that you reject the null hypothesis even though it is true. It is often set at 5% or 1%.
> 7. The actual probability of rejecting the null hypothesis for a test based on the binomial distribution is usually less than the specified significance level.
> 8. There are two main ways of making a decision in hypothesis testing:
>   - Using **probability**. The probability of the observed result or a more extreme result is compared with the significance level. This probability is called the **p-value**; if it is less than the significance level, H_0 is rejected, otherwise it is accepted.
>   - Using the **critical region**. The critical region is the set of values of X for which the probability of an extreme outcome is less than the significance level. If the observed value lies in the critical region, H_0 is rejected, otherwise it is accepted.
> 9. The hypothesis test should end with a non-assertive conclusion in context.

You need to set out your working carefully when testing a hypothesis. This will help you to organise your thinking as well as getting you marks in the examination. The stages in the process are illustrated in the following situation.

The head of a large school announces that to give equal opportunities to all students, the School Council will be chosen at random from all the students, instead of being elected, because an election is biased towards popular students. There are equal numbers of boys and girls in the school, but the School Council chosen consists of 10 girls and 4 boys. Is there evidence (at the 5% level) of a bias towards girls in the selection process?

Deciding what the hypotheses are

If there is no bias, each vacancy is equally likely to be filled by a boy as it is to be filled by a girl. Let p = the probability of a vacancy being filled by a girl. The two possibilities are:
$$p = \frac{1}{2} \text{ and } p > \frac{1}{2}.$$
The first of these assumes the new policy is working so it is the null hypothesis. The hypotheses are:

$$\left.\begin{array}{l} H_0 : p = \frac{1}{2} \\ H_1 : p > \frac{1}{2} \end{array}\right\} \text{where } p \text{ is the probability of a girl being chosen to fill a vacancy.}$$

> You are looking for evidence of a bias towards girls (and so evidence that the policy isn't working).

> **Hints:**
> - The null hypothesis is the belief you start with. You will only stop believing this if there is enough evidence.
> - The alternative hypothesis is what you are looking for evidence of.
> - For a hypothesis test using the binomial distribution, the null hypothesis must be p = a specific value.
> - You must say what p stands for.
> - You will get marks in the exam for writing down the hypotheses correctly even if you go wrong with the rest of the question.

There are 14 vacancies. Assuming the null hypothesis is true, the number of girls on the council, X, would have a binomial distribution.

$$X \sim B(14, 0.5) \text{ if } H_0 \text{ is true.}$$

> The null hypothesis, H_0, says that there is no bias.

Finding a critical region

To find the critical region, it is helpful to have a mental picture of the probability distribution.

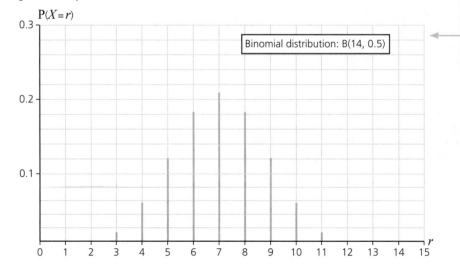

> **Hint:** You are looking for evidence of bias towards girls so need to find numbers of girls which are unusually large.

Using a calculator, $P(X \geqslant 11) = 0.0286...$

Hints:
- The low probability shows that getting 11 or more girls on the School Council would be unusual if there was no bias. 0.0286... is not 5% (the significance level of the test) so you should try looking at a larger 'tail'.
- The level at which you decide 'This is so unlikely that I do not believe the null hypothesis is true after all' is called the **significance level** of the test. In this case it was set at 5%, or 1 in 20. You are usually given the significance level in the question but sometimes you may be asked what it means as well.

Using a calculator, $P(X \geqslant 10) = 0.0897...$

Hint: 0.0897 is bigger than 5%. It is not possible to get exactly 5% so use the 'tail' with a lower probability than 5% rather than the one with a higher probability.

Common mistake: Write down all the probabilities you work out when looking for a critical region. Even though you do not use the probability of 0.0897..., it is important that you worked it out because this is what tells you that the critical region is $X \geqslant 11$.

The critical region is $X \geqslant 11$.

Hint: The critical value is 11; this is the boundary of the critical region.

The number of girls on the School Council was 10. This is not in the critical region so you do not reject the null hypothesis. There is not enough evidence, at the 5% level, to show that there has been any bias towards girls.

Common mistake: You can never be certain whether the null hypothesis was true so you should not make definite claims in your conclusion – state the conclusion in terms of there either being or not being enough evidence of what the alternative hypothesis claimed.

Calculating a *p*-value

The increasing use of software in statistics has led to the use of *p*-values in hypothesis testing becoming more common.

The observed value was $X = 10$ and this raised a doubt as to whether H_0 is true; you would be even more doubtful if there were more girls so find $P(X \geqslant 10)$.

Using a calculator,

$P(X \geqslant 10) = 0.0898$.

The *p*-value is 0.0898; this is more than the significance level. There is not sufficient evidence of a bias towards girls at the 5% level of significance.

Common mistake: The *p*-value is the probability of getting an outcome at least as extreme as that observed, if the null hypothesis is true. Don't confuse it with the parameter *p* which appears in the null and alternative hypotheses.

Common mistake: Do not work out the probability of just one value, $P(X = 10)$ in Example 1, because the probability for any particular value of X can be very small (especially if *n* is large) so you could be in the position of deciding that none of the outcomes are likely, but something has to happen. Always work out the *p*-value by finding probability of the observed outcome *together* with those which cast even more doubt on the null hypothesis.

Hint: Compare the probability you work out with the significance level and reject H_0 if the probability is smaller than the significance level. This is equivalent to supposing that the critical region is $X \geqslant 10$ for Example 1, calculating the significance level and comparing it with the significance level that is required.

Putting it all together (using a *p*-value)

All the stages in a hypothesis test are shown together in the following example. This example uses a *p*-value.

Worked example

Example 1

A sweet-making machine has been producing 20% misshapes. After the machine is serviced, a sample of 30 sweets is taken and 2 of them are misshapes. Is there evidence, at the 5% level of significance, that the machine is producing fewer misshapes?

Hint: It will often be up to you whether you use a *p*-value or a critical region when doing a hypothesis test, but some questions may ask for you to do one of these rather than the other, so you need to be familiar with both methods.

Solution

$H_0 : p = 0.2$ | where p is the proportion of misshapes
$H_1 : p < 0.2$ | produced by the machine.

Remember the null hypothesis starts $p = \ldots$

Remember to say what letters stand for.

You were asked to look for evidence that the machine is producing fewer misshapes.

X = number of misshapes in the sample of 30

If the null hypothesis is true, X has a binomial distribution with $n = 30$, $p = 0.2$.

$X \sim B(30, 0.2)$

Write down the distribution X would have if H_0 were true.

Need to find $P(X \leqslant 2)$.

$X = 2$ is what happened. A smaller value of X would make you believe the alternative hypothesis, that $p < 0.2$, even more.

Using a calculator,

$P(X \leqslant 2) = 0.04418 = 4.42\%$.
This is smaller than 5% so reject H_0.

There is sufficient evidence, at the 5% level of significance, to suggest that the machine is producing fewer misshapes.

State the conclusion non-assertively in the context of the original question.

Common mistakes:

- You know that 2 out of 30 in the sample were misshapes. That is 6.7% of the sample, but you want to know about the percentage of misshapes in all the objects the machine produces. Is it likely to be less than 20%, or have you just got a good sample by chance?

- Statistical hypothesis testing does not allow you to be sure whether the machine is producing fewer misshapes or not; it provides evidence to help you make a decision.

A diagrammatic representation

For the situation in Example 1, the probabilities for the different values that X can take, if the null hypothesis is true, are shown in the vertical line chart on the next page.

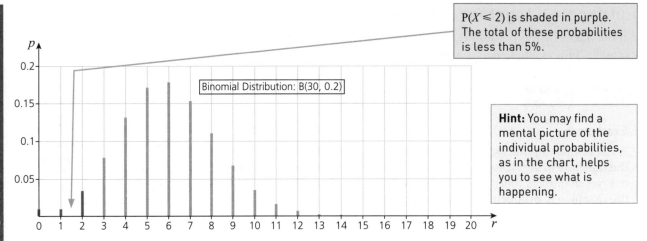

P(X ⩽ 2) is shaded in purple. The total of these probabilities is less than 5%.

Hint: You may find a mental picture of the individual probabilities, as in the chart, helps you to see what is happening.

Putting it all together (using a critical region)

Worked example

Example 2

A phone company knows its customers are dissatisfied with the helpline so it decides to train its staff. First the company surveys its customers and finds that 30% of them are dissatisfied with the helpline. Following training for the staff on the helpline, the company surveys a random sample of 100 customers and finds that 23 of them are dissatisfied with the helpline. It carries out a suitable hypothesis test.

i   State the null and alternative hypotheses.

ii Find the critical region for the relevant test at the 5% level of significance.

iii Use the critical region to decide whether there is evidence that fewer of the customers are dissatisfied with the helpline.

Solution

i   The hypotheses are:

$H_0 : p = 0.3$

$H_1 : p < 0.3$

The company is looking for evidence that fewer customers are dissatisfied.

p = the proportion of customers who are dissatisfied with the helpline.

X = the number of dissatisfied customers in a sample of size 100.

Remember to say what letters stand for.

ii If the null hypothesis is true, X will have a binomial distribution with $n = 100$ and $p = 0.3$.

This is written $X \sim B(100, 0.3)$.

To find the critical region, it is helpful to have a mental picture of the probability distribution.

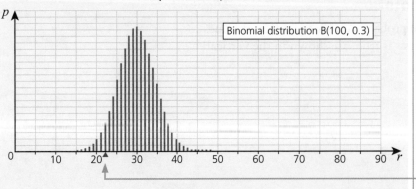

The critical value is 22. The critical region is $X \leqslant 22$.

The null hypothesis will be rejected if there is a small number of dissatisfied customers in the sample. Your task is to find low possible values of X with a total probability of 5%. This is sometimes called the lower tail.

Using a calculator,

$P(X \leq 20) = 0.0164$

$P(X \leq 25) = 0.1631$

$P(X \leq 22) = 0.0479$ ←

$P(X \leq 23) = 0.0755$ ←

The first of these probabilities is 4.79%; the second is 7.55%.

It is not possible to get exactly 5% so go for 4.79%. ←

The critical region is $X \leq 22$.

iii The number of dissatisfied customers in the sample of 100 was 23 so $X = 23$. This is not in the critical region so you do not reject the null hypothesis. There is not enough evidence, at the 5% level, to show that there has been a reduction in the number of dissatisfied customers.

Hint: You need to write down both these probabilities, one each side of 5% so that it is clear how you chose your critical region. You may have to work out other probabilities before you get to two consecutive values, one each side of 5%, but you do not need to write them down.

If the null hypothesis is true, the actual probability of rejecting it for this critical region is 4.79%.

Common mistake: It is almost never possible to get exactly 5% (or whatever the significance level is) when finding a critical region for a binomial hypothesis test. You should always go for a probability lower than the significance level, not the one nearest to it.

Empty critical region

Worked example

Example 3

10% of the population of a country suffer from a particular chronic disease. Following a local public health campaign to reduce levels of the disease, a health worker wants to know whether the campaign has been effective. She decides to take a random sample of 20 people and test them for the disease to see if there is evidence, at the 5% level, that the proportion of people with the disease has dropped. Find the critical region.

Solution

The hypotheses are:

$H_0 : p = 0.1$

$H_1 : p < 0.1$ ←

where p is the proportion of people in the area with the disease

X = the number of people with the disease in a sample of 20

If the null hypothesis, H_0, is true, X has a binomial distribution with $n = 20$ and $p = 0.1$ which can be written $X \sim B(20, 0.1)$.

From a calculator,

$P(X = 0) = 0.1216 = 12.16\%$.

This is more than 5% so the critical region is empty.

She is looking for evidence that the proportion has dropped below 10%.

Hint: The empty critical region means you cannot get enough evidence to reject H_0, at the 5% level, from a sample of 20 people. You need a larger sample for this test.

Test yourself

The following situation is for Questions 1–4. A local authority claims that two-thirds of households in their area recycle glass. An environmentalist thinks that this is based on out-of-date data and that more households now recycle glass. He takes a random sample of 20 households and finds that 16 recycle glass. He wishes to test, at the 5% level of significance, whether there is evidence that more than two-thirds of households in the area recycle glass.

1 Which of these gives the hypotheses for the test which the environmentalist uses?
In each case, p = proportion of households in the area that recycle glass.

A $H_0 : p > \frac{2}{3}$ B $H_0 : p = \frac{2}{3}$ C $H_0 : p = \frac{2}{3}$ D $H_0 : p < \frac{2}{3}$ E $H_0 : p = 0.8$

 $H_1 : p = \frac{2}{3}$ $H_1 : p > \frac{2}{3}$ $H_1 : p \neq \frac{2}{3}$ $H_1 : p > \frac{2}{3}$ $H_1 : p < 0.8$

2 X = the number of households in the sample that recycle glass. Which one of the following probabilities will give the p-value for the hypothesis test?

A $P(X > 16)$ B $P(X = 16)$ C $P(X \geq 16)$ D $P(X < 16)$ E $P(X \leq 16)$

Before doing Question 3, you need to get the correct p-value.

3 Which is the decision the environmentalist should come to, with the correct reason?
A Accept H_0 because the p-value is more than 5%.
B Reject H_0 because the p-value is more than 5%.
C Accept H_0 because the p-value is different from 5%.
D Reject H_0 because the p-value is different from 5%.
E Reject H_0 because $\frac{16}{20} > \frac{2}{3}$.

4 Which of the following is the best statement of the final conclusion to the test?
A Accept H_0 at the 5% significance level.
B Reject H_0.
C It is certain that no more than two thirds of households recycle glass.
D There is insufficient evidence, at the 5% significance level, to justify the claim that more than two thirds of households recycle glass.
E Accept the alternative hypothesis at the 5% significance level.

5 A binomial hypothesis test is conducted with a sample size n, significance level 5% and hypotheses:
$H_0 : p = \frac{1}{3}$

$H_1 : p < \frac{1}{3}$

X is the number of successes. Four of the following statements about critical regions are true and one is false. Find the one that is false.
A There is only one value in the critical region for $n = 12$.
B If $n = 18$, the critical region is the same as it would be for a significance level of 10%.
C $n = 8$ is the smallest value of n for which the critical region is non-empty.
D For $n = 16$, the critical region is $X \leq 2$.
E The critical region is the same for each of $n = 8$, 9, 10, 11, 12.

Full worked solutions online

CHECKED ANSWERS

Exam-style question

A report from a few years ago says that 90% of children (aged 11–16) in an area have mobile phones. A researcher suspects that the percentage with mobile phones is greater now. She asks a random sample of 200 children aged 11–16 from the area and finds that 193 have mobile phones. A suitable statistical test will be carried out to see whether there is evidence that the proportion with phones has increased.
i Write down suitable null and alternative hypotheses.
ii Carry out the test at the 5% level of significance, stating your conclusions clearly.
iii Would your conclusions be different at the 1% level of significance? Explain your reasoning.

Short answers on page 223

Full worked solutions online

CHECKED ANSWERS

Key facts

1   For a hypothesis test using the binomial distribution, the null hypothesis is written in the form $H_0: p = 0.3$ (or some other specific value).

2   For a 1-tail test the alternative hypothesis that goes with the example null hypothesis above is EITHER $H_1: p > 0.3$ OR $H_1: p < 0.3$. 1-tail tests were covered in the previous section.

3   For a 2-tail test the alternative hypothesis that goes with the example null hypothesis above is $H_1: p \neq 0.3$ (this includes both possibilities: $p > 0.3$ and $p < 0.3$). 2-tail tests are covered in this section.

4   2-tail tests can be carried out using either probability or critical regions.

5   The significance level for a 2-tail test is split into two halves: one half for each tail.

Symmetrical and asymmetrical binomial distributions

The binomial distribution is symmetrical if $p = 0.5$. It is skewed for other values of p. This is illustrated in the vertical line graphs below.

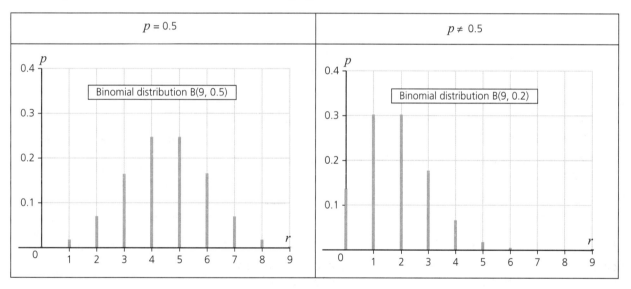

Critical region for a symmetrical 2-tail test

Worked example

Example 1

As part of an engineering project, Alicia invents a machine that is intended to replace tossing a coin at the start of a sports match. The machine should show 'heads' or 'tails' with equal probability. To test it for bias, she will run it 20 times. For what numbers of 'heads' will she conclude that it is biased, at the 5% level of significance?

Solution

$H_0 : p = 0.5$

$H_1 : p \neq 0.5$

where p is the probability of 'heads' on each run.

If the null hypothesis is true and the probability of 'heads' is, indeed, 0.5 on each go then the number of heads in 20 runs will have a binomial distribution with $n = 20$, $p = 0.5$. This can be written as follows:

If H_0 is true, $X \sim B(20, 0.5)$, where X is the number of 'heads' on 20 runs.

For a 2-tailed critical region at the 5% level of significance, you need 2.5% for each tail. The two tails are coloured in purple on the vertical line chart below.

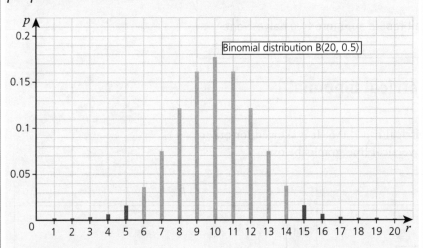

Binomial distribution B(20, 0.5)

For the lower tail, look for a probability of 2.5%; it is unusual to be able to find exactly 2.5% so write down the one below 2.5% and the one above 2.5%.

Using cumulative binomial probabilities on a calculator gives the following:

$P(X \leqslant 5) = 0.0207 = 2.07\%$

$P(X \leqslant 6) = 0.0577 = 5.77\%$

The critical region for the lower tail is $X \leqslant 5$.

For the upper tail, use cumulative binomial probabilities to look for probabilities of the form $P(X \geqslant x)$ which are close to 2.5%.

$P(X \geqslant 16) = 0.0059$

$P(X \geqslant 15) = 0.0207$

$P(X \geqslant 14) = 0.0577$

The upper tail consists of the high values with a total probability of 2.5% or less, so the critical region for the upper tail is $X \geqslant 15$.

She should conclude the machine is biased if $X \leqslant 5$ or $X \geqslant 15$, where X is the number of 'heads' on 20 runs.

If the machine is unbiased, the probability of 'heads' on each go is 0.5.

She is looking for evidence of bias in either direction.

Remember to say what X and p stand for.

Hint: You need to write down both these probabilities so that it is clear how you chose your critical region.

Hint: Choose the one with a probability below 2.5% rather than the one with a probability above 2.5%.

Hint: You will need to either use $P(X \geqslant x) = 1 - P(X \leqslant (x-1))$ or enter lower and upper limits. Remember $n = 20$, $p = 0.5$.

Hints:

- If H_0 is $p = 0.5$, the critical regions for the upper and lower tails are symmetrical.

- For a 2-tailed critical region, work out half the significance level and then carry on as if you were working out two separate critical regions: one for $H_1: p < 0.5$ and one for $H_1: p > 0.5$.

If the null hypothesis is true, the actual probability of rejecting it for this critical region is 2.07% + 2.07% = 4.14%.

Critical region for an asymmetrical 2-tail test

Worked example

Example 2

A seed firm sells a particular variety of flower seeds which have an 85% germination rate. The seeds are expensive, so they are stored in carefully controlled conditions. A batch of the seeds has been stored in different conditions and a sample of 25 of this batch is planted to see if the rate of germination has been affected.

i What numbers of germinating seeds would lead to the conclusion, at the 10% significance level, that there has been a change in the germination rate?

ii Hence, decide whether 23 seeds germinating would provide evidence of such a change.

Solution

i $H_o : p = 0.85$

   $H_1 : p \neq 0.85$

where p = proportion of the seeds in the batch that will germinate.

If the null hypothesis is true and the proportion of seeds in the batch germinating is, indeed, 0.85 then the number of germinating seeds from a sample of 25 will have a binomial distribution with $n = 25$, $p = 0.85$.

This can be written as follows:

If H_o is true, $X \sim B(25, 0.85)$, where X is the number of seeds in the sample of 25 that germinate.

For a 2-tailed critical region at the 10% level of significance, you need 5% for each tail.

Using cumulative binomial probabilities on a calculator (with $n = 25$, $p = 0.85$) gives the following.

$P(X \leqslant 17) = 0.0255 = 2.55\%$ ◄———

$P(X \leqslant 18) = 0.0695 = 6.95\%$ ◄———

> **Hint:** Write down the probability below 5% and the one above 5%.

The critical region for the lower tail is $X \geqslant 17$. ◄———

> **Hint:** Choose the one with a probability below 5% rather than the one with a probability above 5%.

For the upper tail, use cumulative binomial probabilities to look for probabilities of the form $P(X \geqslant x)$ which are close to 5%.

$P(X \geqslant 24) = 0.0931$

$P(X \geqslant 25) = 0.0172$

> **Hint:** You will need either to use $P(X \geqslant x) = 1 - P(X \leqslant (x-1))$ or to enter lower and upper limits. Remember $n = 25$, $p = 0.85$.

The critical region for the upper tail is $X \geqslant 25$. This simplifies to $X = 25$ as only 25 seeds were planted so no more than 25 can germinate.

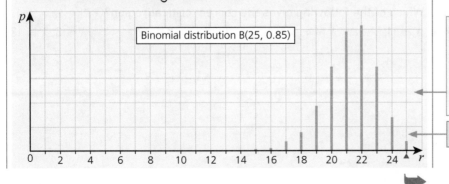

Binomial distribution B(25, 0.85)

> **Hint:** You need not spend time drawing a vertical line chart of the binomial distribution but imagining it, or drawing a rough sketch, may help you understand the process.

> The critical value is $X = 25$.

> If 17 or fewer seeds germinate or if 25 seeds germinate, this will provide evidence, at the 10% level, that the germination rate has changed.
>
> ii 23 is not in the critical region so it does not provide evidence of a change at the 10% level.

Common mistake: Notice that the two parts of the critical region are not symmetrical.

Using probability to conduct a 2-tail hypothesis test

It is possible to conduct a 2-tail test by working out the probability of the 'tail'. It is not always obvious for an asymmetrical distribution whether a value lies in the upper or lower tail so work out the expected value first to help you decide.

The process is illustrated using the information from Example 2 above.

Expected value $= np = 25 \times 0.85 = 21.25$

The observed value was 23; this is bigger than 21.25 so you want to look at the upper tail.

Now work out the tail probability that includes 23; this is $P(X \geqslant 23) = 0.2537 = 25.37\%$

Since the test is a 2-tail test, you should compare this with half the significance level.

25.37% > 5% so there is insufficient evidence of a change at the 10% level.

Common mistake: It may seem obvious that 23 would be in the upper tail rather than the lower tail with $n = 25$ but it is worth working out the expected value. A value such as 20 would be in the lower tail because it is below the expected value and you might not realise this unless you worked out the expected value.

Hint: Remember to use half the significance level for each tail because this is a 2-tail test.

Common mistake: 0.2537 is the probability of 23 or more but it is not the p-value for the test because it only refers to one tail and this is a 2-tail test. There are different approaches to working out p-values for 2-tail tests for asymmetrical distributions; you are not expected to know them.

Test yourself

TESTED

For Questions 1 and 2, X is the number of successes for a binomial random variable and p is the probability of success on each trial.

1 A sample of size 19 is taken to test the hypotheses

$H_0 : p = 0.5$

$H_1 : p \neq 0.5$

at the 1% level of significance. Which of the following gives the correct critical region?

A $X \leqslant 3, X \geqslant 15$ B $X \leqslant 3, X \geqslant 16$ C $X \leqslant 4, X \geqslant 15$ D $X \leqslant 5, X \geqslant 14$ E $X \leqslant 5, X \geqslant 16$

2 A sample of size 14 is taken to test the hypotheses

$H_0 : p = 0.6$

$H_1 : p \neq 0.6$

at the 10% level of significance. Which of the following gives the correct critical region?

A $X \leqslant 4, X \geqslant 10$ B $X \leqslant 4, X \geqslant 11$ C $X \leqslant 4, X \geqslant 12$ D $X \leqslant 5, X \geqslant 12$ E $X \leqslant 3, X \geqslant 13$

The following information will be used in Questions 3 and 4:

As part of a psychology experiment, a random sample of 18 people was asked to choose one of two puzzles: either a normal sudoku, using numbers, or a puzzle which used letters instead of numbers, but which was otherwise identical to the first one. The experimenter wants to test whether there is a bias towards choosing either letters or numbers. p is the proportion of people in the population who would choose the puzzle with numbers. 13 people in the sample chose the puzzle with numbers. The test is conducted at the 5% level of significance.

3 Which of the following are the hypotheses for the test?

A	$H_0 : p > 0.5$	B	$H_0 : p > 0.5$	C	$H_0 : p = 0.5$	D	$H_0 : p = 0.5$	E	$H_0 : p \neq 0.5$
	$H_1 : p < 0.5$		$H_1 : p = 0.5$		$H_1 : p > 0.5$		$H_1 : p \neq 0.5$		$H_1 : p = 0.5$

Make sure you understand which are the correct hypotheses before attempting Question 4.

4 Which of the following is the correct conclusion for the test?

A The critical region is $X \leqslant 4, X \geqslant 13$ so there is evidence of a bias.
B The critical region is $X \leqslant 4, X \geqslant 14$ so there is not sufficient evidence of a bias.
C The critical region is $X \leqslant 4, X \geqslant 14$ so there is evidence of a bias.
D The critical region is $4 \leqslant X \leqslant 14$ so there is not sufficient evidence of a bias.
E The critical region is $X \leqslant 5, X \geqslant 13$ so there is evidence of a bias.

5 A teacher reads an article which says that 15% of the population are left handed. She wonders whether the proportion is the same among students taking A Level Mathematics.
She observes 19 students taking a mathematics examination and notes how many of them are left handed (X). Assuming these 19 students are a random sample from the population of A Level Mathematics students, what is the critical region for the test at the 5% level of significance?

A	$X \geqslant 4$	B	The critical region is empty
C	$X \geqslant 6$	D	$X = 0, X \geqslant 7$
E	$X \geqslant 7$		

Full worked solutions online

CHECKED ANSWERS

Exam-style question

A plate factory has historically had 5% of the plates produced being faulty. Following changes to production methods, the management wants to investigate whether the proportion of faulty plates has changed. A random sample of 40 plates contains 1 faulty plate.

i   Write down suitable hypotheses.
ii What is the critical region for a suitable hypothesis test, at the 5% level?
iii Carry out the test, stating your conclusions clearly.
iv If the null hypothesis is true, what is the probability of the test statistic being in the critical region?

Short answers on page 223

Full worked solutions online

CHECKED ANSWERS

Chapter 6 Discrete random variables and the Normal distribution

About this topic

A random variable can take different values. A discrete random variable can only take discrete values. Examples of discrete random variables include the total score when two dice are thrown, the proportion of heads when tossing 4 coins, the number of children in a family. The number of successes in a binomial distribution is an example of a discrete random variable. The probability distribution for a discrete random variable gives the probability of each possible value.

The Normal probability distribution is an example of a continuous random variable; it is used to describe measurements of many naturally occurring variables such as human height or leaf length. It can also be used as an approximating distribution. If you study or work in medical, biological or social sciences you will use the Normal distribution. There are an infinite number of possible values of a continuous random variable so probabilities are used for ranges of possible values.

Before you start, remember ...

- Basic ideas of probability from GCSE.
- Types of data and vertical line charts from chapter 2.
- Substituting into formulae from GCSE.
- How to solve equations and simultaneous equations from GCSE.

Discrete random variables

REVISED

Key facts

1. Upper case (capital) letters are used to stand for discrete random variables; e.g. S could stand for 'the total score when two dice are thrown'.

2. Lower case letters are used to stand for values of the discrete random variable; e.g. $s = 2, 3, 4, ..., 12$.

3. Each possible value of the random variable has a probability of occurring which is between 0 and 1.

4. If the random variable X can take values $r_1, r_2, r_3, ..., r_n$ with

   probabilities $p_1, p_2, p_3, ..., p_n$ respectively then $\sum_{i=1}^{n} p_i = 1$, i.e. all the probabilities add up to 1.

5. In a discrete uniform distribution, each possible value has the same probability.

6. A discrete probability distribution can be illustrated by a vertical line chart.

Discrete random variables

Discrete random variables can only take discrete values. Each possible value has a probability of occurring.

Worked example

Example 1

Two fair four-faced dice are each numbered 1 to 4. Both dice are thrown and the two scores are added. Find the probability of each possible total.

Solution

The probabilities can be found by drawing a sample space diagram (shown below).

Total	1	2	3	4
1	2	3	4	5
2	3	4	5	6
3	4	5	6	7
4	5	6	7	8

Total	2	3	4	5	6	7	8
Probability	$\frac{1}{16}$	$\frac{2}{16}$	$\frac{3}{16}$	$\frac{4}{16}$	$\frac{3}{16}$	$\frac{2}{16}$	$\frac{1}{16}$

Hint: There are 16 items in the sample space, three of them are 'total = 4'.

Discrete uniform distribution

For each of the dice, the scores follow a discrete uniform distribution, because each score is equally likely.

Score	1	2	3	4
Probability	$\frac{1}{4}$	$\frac{1}{4}$	$\frac{1}{4}$	$\frac{1}{4}$

The probability distribution of a discrete random variable

The probability distribution can be a table which gives a probability for each value (as in Example 1 above) or it can be a formula which gives a probability for each value (as in Example 2 below).

- For each possible value of the random variable, its probability is between 0 and 1.
- If the random variable X can take values $r_1, r_2, r_3, ..., r_n$ with probabilities $p_1, p_2, p_3, ..., p_n$ respectively then $\sum_{i=1}^{n} p_i = 1$. That is, all the probabilities add up to 1.

Worked example

Example 2

The random variable, X, can take values 1, 2, 3, 4. The probability of each possible value is given by $P(X = r) = \frac{r(5-r)}{20}$.
Find $P(X = 2)$.

Hint: Put 2 in the formula in place of r.

Solution

$$P(X = 2) = \frac{2 \times 3}{20} = \frac{6}{20} = \frac{3}{10}.$$

Worked example

Example 3

The random variable X has a probability distribution given by the formula $P(X = r) = k(6 - r)$ for $r = 1, 2, 3$. k is a constant. Find the value of k.

Solution

r	1	2	3
$P(X = r)$	5k	4k	3k

The probabilities add up to 1 so $5k + 4k + 3k = 1$

$$12k = 1$$

$$k = \frac{1}{12}$$

> **Hint:** Start by putting the values of r into the formula and writing the probabilities in a table.

> **Hint:** Knowing that all the probabilities add up to 1 will be useful for answering exam questions, but it is also a useful check when working out all the probabilities in a probability distribution.

Illustrating discrete random variables

A vertical line chart is a good way to illustrate the probability distribution of a discrete random variable. The following line chart illustrates the probability distribution in Example 3.

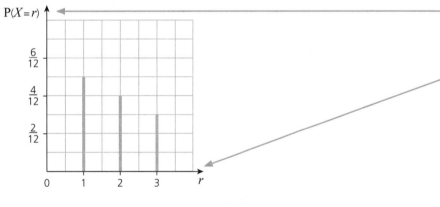

> The probability is on the vertical axis.

> The horizontal axis shows the possible values.

> **Hint:** Notice that the vertical scale does not need to go above 1. Sometimes it is easier to number it using fractions rather than decimals.

Working with discrete random variables

You can use the probabilities for discrete random variables in all the usual ways. Different formulae can be used for the probabilities for different values.

Worked example

Example 4

The probability distribution of the random variable X is given by

$$P(X = r) = \frac{r(r+2)}{100} \quad \text{for } r = 1, 2, 3, 4, 5$$

$$P(X = r) = k \quad \text{for } r = 6 \text{ (k is a constant)}$$

$$P(X = r) = 0 \quad \text{otherwise}$$

> This is just a way of saying that values of r which are not 1, 2, 3, 4, 5, 6 cannot happen.

i Find the value of k.

ii Two independent values of X are generated, one after the other. Find the probability that their total is greater than 10.

Solution

i Using the formula $P(X = r) = \dfrac{r(r + 2)}{100}$, for values of r from 1 to 5, gives the probabilities in this table:

r	1	2	3	4	5	6
$P(X = r)$	0.03	0.08	0.15	0.24	0.35	k

The probability for 6 is just k.

$0.03 + 0.08 + 0.15 + 0.24 + 0.35 + k = 1$

Hint: All the probabilities add up to 1.

$0.85 + k = 1$ so $k = 0.15$

ii A total greater than 10 can come from these pairs of values (5, 6) or (6, 5) or (6, 6).

Common mistake: 5 then 6 is different to 6 then 5, so both need to be included.

$P(\text{total} > 10) = P(5, 6) + P(6, 5) + P(6, 6)$

$P(\text{total} > 10) = 0.35 \times 0.15 + 0.15 \times 0.35 + 0.15 \times 0.15$

$= 0.1275$

Hints:

- The two values are independent, so to work out the probability of 5 and 6 you multiply the two probabilities.
- You could draw a tree diagram for part ii using the probabilities from part i and you should do so if it will help you to see how to work it out.

Modelling with discrete random variables

Theoretical probability distributions are used as approximations to real life situations so that predictions can be made about what is likely to happen.

Worked example

Example 5

The number of eggs laid by a hen in a week, X, is modelled by the probability distribution $P(X = r) = k(7.5 - r)(r - 2.5)$ for $r = 3, 4, 5, 6, 7$. k is a constant.

i Find the proportion of weeks in which, according to the model, the hen lays 7 eggs.

ii A hen is likely to lay 150 to 200 eggs a year with more eggs in the summer than in the winter. Is the probability distribution suitable for modelling the number of eggs laid in winter? Justify your answer.

Hint: You need to work out the value of k so start by putting the values of r into the formula and writing the probabilities in a table.

Solution

i

r	3	4	5	6	7
$P(X = r)$	2.25k	5.25k	6.25k	5.25k	2.25k

$2.25k + 5.25k + 6.25k + 5.25k + 2.25k = 1$

Remember all the probabilities add up to 1.

$21.25k = 1$

$k = 1 \div 21.25 = \dfrac{4}{85}$

$P(X = 7) = 2.25k = 2.25 \times \dfrac{4}{85} = \dfrac{9}{85} \approx 0.106$

Hint: The final answer is more easily understood as a rounded decimal or percentage than as an exact fraction; the match between the model and reality is not likely to be exact.

The hen lays 7 eggs in about 10% of weeks.

ii 150 to 200 eggs a year is, on average, 3 or 4 eggs a week over the whole year. There will be fewer eggs in winter but the model has over 4 eggs with probability $13.75k \approx 0.65$ so it is unlikely to be a good model in winter.

> The main thing in part ii is to present a well-argued reason for your conclusion as to whether the model is suitable for eggs laid in winter.

Test yourself

1 One of the five options in this question could be the probability distribution of a discrete random variable. The other four are not. Find the one which is.

A

r	0	1	2	3
$P(X=r)$	0.56	k	0.24	0.21

B $P(X=r) = k(r+1)$ for $r = 0, 1, 2$

$P(X=r) = kr^2$ for $r = 4, 5$

$P(X=r) = 0$ otherwise

C $P(X=r) = k(r^2 - 2r)$ for $r = 1, 2, 3$

D

r	2	3	4	5
$P(X=r)$	0.36	0.18	0.23	0.21

E $P(X=r) = \dfrac{kr}{r+1}$ for integer values of r between 1 and 5 (inclusive), i.e. $1 \leqslant r \leqslant 5$

2 A discrete random variable, X, has the probability distribution:

$P(X=r) = k(2r^2 - r)$ for $r = 1, 2, 3$

$P(X=r) = 3k$ for $r = 4$

$P(X=r) = 0$ otherwise

Find the value of k.

A 25   B $\dfrac{1}{22}$ C $\dfrac{1}{54}$ D 1   E 0.04

3 A discrete random variable, X, has the probability distribution shown in the table.

r	0	1	2	3	4
$P(X=r)$	0.25	0.2	0.15	0.3	0.1

Find $P(1 \leqslant X < 3)$.

A 0.15   B 0.35   C 0.45   D 0.55   E 0.65

4 X = the larger score when two fair dice are thrown. Which of the following is the correct probability distribution for X?

A $P(X=r) = \dfrac{2r-1}{36}$ for $r = 1, 2, 3, 4, 5, 6$

B $P(X=r) = \dfrac{1}{6}$ for $r = 1, 2, 3, 4, 5, 6$

C $P(X=r) = \dfrac{2r+1}{36}$ for $r = 1, 2, 3, 4, 5, 6$

D $P(X=r) = \dfrac{r}{21}$ for $r = 1, 2, 3, 4, 5, 6$

E $P(X=r) = \dfrac{(r-1)}{15}$ for $r = 1, 2, 3, 4, 5, 6$

5 Barry Mitchell is a footballer who has a 0.75 chance of scoring a goal from a penalty. He takes 2 penalties. Assume that these are independent of each other. Which of the following could be the probability distribution of the number of goals he scores?

A

No of goals	0	1	2
Probability	$\frac{1}{16}$	$\frac{3}{16}$	$\frac{9}{16}$

B

No of goals	0	1	2
Probability	$\frac{1}{16}$	$\frac{3}{16}$	$\frac{12}{16}$

C

No of goals	0	1	2
Probability	0	0.75	0.5625

D

No of goals	0	1	2
Probability	$\frac{1}{3}$	$\frac{1}{3}$	$\frac{1}{3}$

E

No of goals	0	1	2
Probability	$\frac{1}{16}$	$\frac{3}{8}$	$\frac{9}{16}$

Full worked solutions online

CHECKED ANSWERS ☐

Exam-style question

A spreadsheet shows four random integers, each between 1 and 4 inclusive, in a row. X is the number of random digits needed (starting from the left) to get a total of 4 or more. For example, when the random digits are as shown below, $X = 4$ because $1 + 1 + 1 = 3$ and $1+1+1+2=5$.

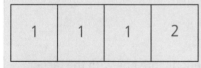

1	1	1	2

The probability distribution for X is shown in the table below.

r	1	2	3	4
$P(X = r)$	$\frac{1}{4}$	$\frac{9}{16}$	p	$\frac{1}{64}$

i Find the value of p.

ii Show that $P(X = 1) = \frac{1}{4}$.

iii Show that $P(X = 2) = \frac{9}{16}$.

Short answers on page 223

Full worked solutions online

CHECKED ANSWERS ☐

The Normal distribution

Key facts

1   If a random variable, X, has a Normal distribution with mean μ and standard deviation σ, this can be written as $X \sim N(\mu, \sigma^2)$.

2   The standard Normal distribution has mean 0 and standard deviation 1. A variable with this distribution is often given the symbol Z. $Z \sim N(0, 1)$. Other Normal variables can be standardised by subtracting the mean and then dividing by the standard deviation. $z = \dfrac{x - \mu}{\sigma}$.

3   The peak of the Normal curve is at the mean. It has a point of inflection one standard deviation from the mean.

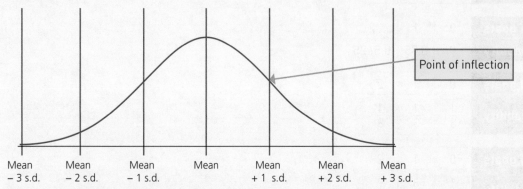

Point of inflection

| Mean
− 3 s.d. | Mean
− 2 s.d. | Mean
− 1 s.d. | Mean | Mean
+ 1 s.d. | Mean
+ 2 s.d. | Mean
+ 3 s.d. |

4   The area under a Normal curve between two values on the horizontal axis gives the probability of the Normal variable being between those values. The total area under the curve is 1.

5   For a Normal distribution, approximately:
   • 68% of values (about two thirds) lie within one standard deviation of the mean
   • 95% of values lie within two standard deviations of the mean
   • 99.8% of values (nearly all) lie within three standard deviations of the mean.

6   Distributions with different means but the same standard deviation will be translations of each other.

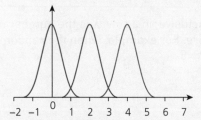

7   For two distributions with the same mean but different standard deviations, the one with the larger standard deviation has a wider curve; the curve is also lower so that the total area under it is still 1.

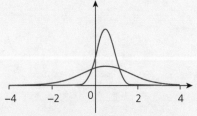

8   The Normal curve is symmetrical and bell-shaped. A distribution which does not have this shape cannot be modelled by a Normal distribution.

9   The Normal distribution is for a continuous variable. If it is used as an approximating distribution for a discrete variable, a continuity correction must be used. For example, values from 12.5 up to 13.5 for the approximating Normal would correspond to a value of 13 for the discrete variable.

What is the Normal distribution?

The Normal distribution is a theoretical distribution which can be used for many naturally occurring variables. $X \sim N(\mu, \sigma^2)$ means that the random variable X has a Normal distribution with mean μ and standard deviation σ (variance σ^2). A Normal distribution has a symmetrical bell-shaped curve; this can be thought of as the limit of histograms showing probabilities for the distribution with narrower and narrower bars.

The area under a Normal curve represents probability so the total area is 1. The area under the curve between two values of x gives the probability that the random variable lies between those values.

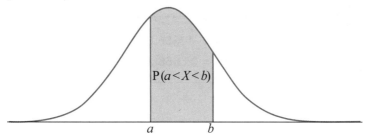

The standard Normal distribution, N(0, 1), has mean 0 and variance 1.

Using calculators to find probabilities

Calculators suitable for A Level Mathematics will give probabilities from Normal distributions; typically they work in the way illustrated in Example 1. You need to know how to use your calculator to find a probability for any Normal distribution.

Worked example

Example 1

Adult men in the UK have a mean height of 175 cm. The standard deviation is 7.5 cm. You can assume that the heights are Normally distributed. Find the probability that a randomly chosen man is shorter than 180 cm.

Solution

What you need to write down	Commentary	What to do on your calculator
H = height of a randomly chosen adult man in cm.	Say what the letters you introduce stand for.	
H ~ N(175, 7.5²)	Write down the probability distribution. Leave the variance as 7.5² because you will need the standard deviation.	
P(H < 180)	This is the probability you were asked to find.	
	Select Normal CD (this gives cumulative probabilities).	1 : Normal PD 2 : Normal CD 3 : Inverse Normal 4 : Binomial PD

	Enter the mean, standard deviation and lower and upper limits. There is no lower limit for the probability you are finding here so use a negative number which is large in size.	Normal C.D Data : Variable Lower : -1E+05 Upper : 180 σ : 7.5 μ : 175 Save Res : None
	On some calculators, you will get the standardised values of the upper and lower limits as well as the probability. There is more about standardised values later in this chapter.	Normal C.D P = 0.74750746 z : Low = -13356.667 z : Up = 0.66666666
The probability is 0.7475 The probability of a randomly chosen man being shorter than 180 cm is 0.748 (to 3 d.p.)		

Hint: For the binomial distribution, it makes a difference whether you are working out $P(X>3)$ or $P(X \geqslant 3)$. This is because it is a discrete distribution and the probabilities for these ranges are found by adding up individual probabilities so whether $P(X=3)$ is included makes a difference. The Normal distribution is continuous and the total area is given by the area under the curve. Whether the line at $X=3$ is included or not, it makes no difference to the area. So, for Example 1 above, $P(H<180)$ is the same as $P(H \leqslant 180)$.

Using inverse Normal probabilities on your calculator

You can use your calculator in the situation shown in the diagram below. You want to find the unknown upper bound, x, of the Normal variable that corresponds to a given cumulative probability, p.

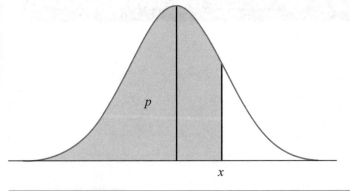

> The probability that the random variable is less than a particular value is called the probability of the lower tail. Some calculators will also work out inverse probabilities for an upper tail but you can always work with the lower tail because the total probability is 1.

Worked example

Example 2

Adult men in the UK have a mean height of 175 cm. The standard deviation is 7.5 cm. You can assume that the heights are Normally distributed.

i   Find the upper quartile.

ii Find the lower quartile.

Solution

What you need to write down	Commentary	What to do on your calculator
H = height of a randomly chosen adult man in cm.	Say what the letters you introduce stand for.	
$H \sim N(175, 7.5^2)$	Write down the probability distribution. Leave the variance as 7.5^2 because you will need the standard deviation.	
i $P(H < h) = 0.75$, where h is the upper quartile.	The probability of a man being shorter than the upper quartile is 0.75.	
	Draw a sketch of the Normal curve. Remember that the line of symmetry is at the mean and the area each side of the line of symmetry is 0.5. Shading the probability you have as an area will help you check you have the right answer.	
	Choose the inverse Normal.	1: Normalpdf 2: Normalcdf **3:** invNormal
	Enter the tail probability, mean and standard deviation.	invNormal area = 0.75 mean = mu = 175 sigma = 7.5
The upper quartile is 180.1 cm.	Check that your answer is sensible; the value is above 175, as expected. Round it sensibly – one or two decimal places is enough for height.	invNormal VALUE = 180.0586731 STORE: ■ xyztabcd
ii $P(H < L) = 0.25$, where L is the lower quartile.	The probability of a man being shorter than the lower quartile is 0.25.	

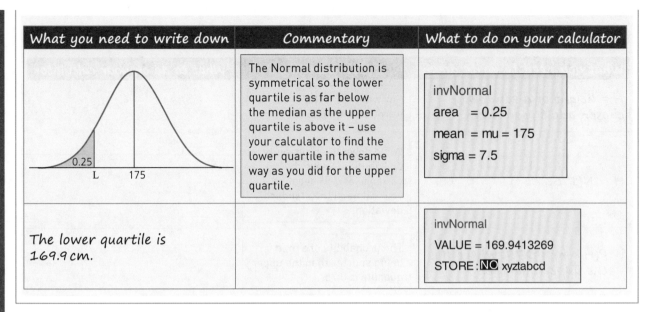

What you need to write down	Commentary	What to do on your calculator
	The Normal distribution is symmetrical so the lower quartile is as far below the median as the upper quartile is above it – use your calculator to find the lower quartile in the same way as you did for the upper quartile.	invNormal area = 0.25 mean = mu = 175 sigma = 7.5
The lower quartile is 169.9 cm.		invNormal VALUE = 169.9413269 STORE: NO xyztabcd

Standardising a Normal distribution

For any Normal distribution, the process of subtracting the mean and then dividing by the standard deviation is called standardising the distribution.

For $X \sim N(\mu, \sigma^2)$, $z = \dfrac{x-\mu}{\sigma}$ is the standardised value. It is the number of standard deviations the particular value of X is from the mean. The values of Z have a Normal distribution with mean 0 and standard deviation 1; this is called the standard Normal distribution.

Finding mean and/or standard deviation from probability

Sometimes you will know that data are Normally distributed but not know the mean and standard deviation. If you know the probability of data values being in a particular range you can work out the mean or standard deviation or both. To do this, you will need to use the standard Normal distribution.

Worked example

Example 3

Heights of adult men in China are Normally distributed. 12.7% of men are shorter than 160 cm and 21.2% of men are taller than 174 cm. Find the mean and standard deviation of the heights of adult men in China.

Solution

M stands for the height, in cm, of a randomly chosen Chinese man. The population mean is μcm and the standard deviation is σcm.

Say what the letters you introduce stand for.

$M \sim N(\mu, \sigma^2)$

12.7% of men are shorter than 160 cm.

$P(M < 160) = 0.127$

Standardising, $P\left(Z < \dfrac{160 - \mu}{\sigma}\right) = 0.127$

$P(Z < z_1) = 0.127$

Hint: You don't know the mean or standard deviation of the distribution so could not use your calculator without standardising.

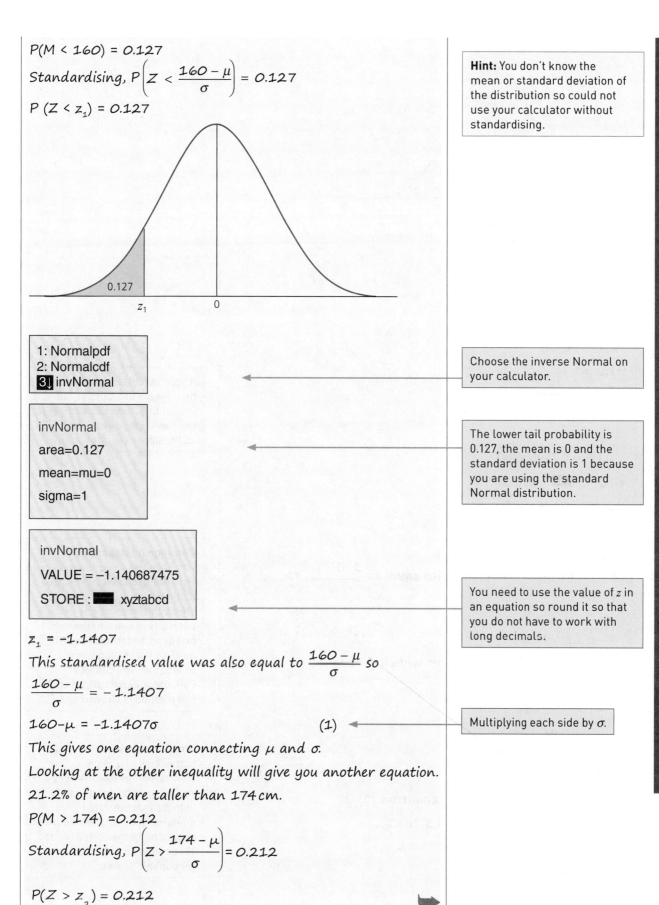

Choose the inverse Normal on your calculator.

The lower tail probability is 0.127, the mean is 0 and the standard deviation is 1 because you are using the standard Normal distribution.

You need to use the value of z in an equation so round it so that you do not have to work with long decimals.

$z_1 = -1.1407$

This standardised value was also equal to $\dfrac{160 - \mu}{\sigma}$ so

$\dfrac{160 - \mu}{\sigma} = -1.1407$

$160 - \mu = -1.1407\sigma$ \hfill (1)

Multiplying each side by σ.

This gives one equation connecting μ and σ.

Looking at the other inequality will give you another equation.

21.2% of men are taller than 174 cm.

$P(M > 174) = 0.212$

Standardising, $P\left(Z > \dfrac{174 - \mu}{\sigma}\right) = 0.212$

$P(Z > z_2) = 0.212$

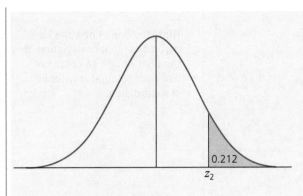

$P(Z < z_2) = 1 - 0.212$
$= 0.788$

invNormal

area = 0.788

mean = mu = 0

sigma = 1

invNormal

VALUE = 0.799500947

STORE : ■ xyztabcd

SOLVE AGAIN QUIT

If your calculator will give the inverse of an upper tail probability, then you can get to the same value of z_2 using 0.212 as the upper tail probability.

$z_2 = 0.7995$

The standardised value was also equal to $\dfrac{174 - \mu}{\sigma}$, so

$\dfrac{174 - \mu}{\sigma} = 0.7995$

$174 - \mu = 0.7995\sigma$ (2)

If you subtract (1) from (2) this will eliminate μ.

$174 - \mu = 0.7995\sigma$

$160 - \mu = -1.141\sigma$

$\overline{14 \qquad = 1.9405\sigma}$

$\sigma = 14 \div 1.9405 = 7.2146.....$

$\sigma = 7.215$

Substitute the value of σ into equation (2).

$174 - \mu = 0.7995 \times 7.2146..... = 5.7681....$

$\mu = 174 - 5.7681.... = 168.231...$

$\mu = 168.23$

Common mistake: Now you have two simultaneous equations to solve. If you are doing this in an examination check the instructions carefully to see whether you are expected to show a full method for solving them or whether you can use a calculator equation solver. The full method is shown here.

Hint: There are a lot of stages to this kind of question. Think about whether the answers you are getting are sensible. In this case, the data in the question suggest that the mean should be between 160 cm and 174 cm and so 168.23 cm is a reasonable answer.

When is the Normal distribution used?

The Normal distribution provides a good model for many naturally occurring variables. The diastolic blood pressures of a sample of 100 adults are shown in the following histogram. The Normal curve on the diagram has the same mean and standard deviation as the data; the Normal curve has been stretched so that the area under it equals the total frequency. It seems reasonable to assume that the Normal model would be a reasonably good match for the blood pressures of the whole adult population.

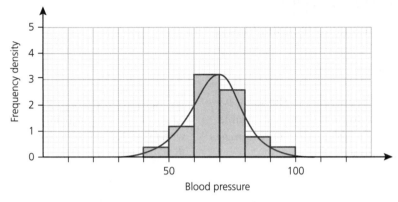

Worked example

Example 4

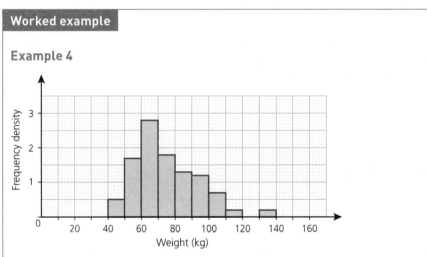

The histogram shows the distribution of weights of patients in a hospital. Explain why the Normal distribution would not be appropriate for this population.

Solution

The distribution of weights is not symmetrical but the Normal distribution is symmetrical so it cannot be used as a model for this distribution of weights.

Common mistake: The Normal distribution is symmetrical and bell-shaped; if the histogram from the data is a different shape then this suggests that the Normal distribution may not be suitable. However, the data are taken from a sample and different samples may give a different impression. You should be cautious when drawing conclusions, especially from small samples.

Modelling discrete situations

IQ tests give a whole number score. They are designed to follow a Normal distribution with the mean for the whole population being 100 and the population standard deviation 15. The distribution of scores is shown in the following vertical line chart.

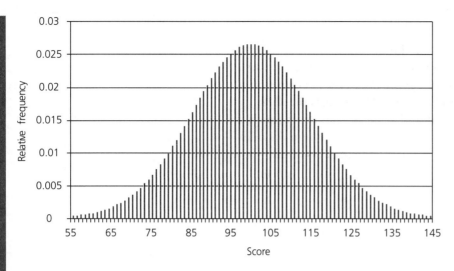

The scores which candidates receive can only be integers so the score is a discrete variable but the Normal distribution is for a continuous variable. To use the Normal distribution for a discrete variable, a *continuity correction* has to be used. An example of this is given below.

> It is the 'unrounded' score which is Normally distributed with mean 100 and standard deviation 15.

Worked example

Example 5

An IQ test is designed so that the scores follow a Normal distribution with a mean of 100 and a standard deviation of 15. Candidates can only get integer scores. Find the proportion of people taking the test who score between 110 and 115 marks (inclusive).

Solution

It will help you to think of the integer score (X) which candidates get and the 'unrounded' score (S), which is Normally distributed.

Scores from 110 to 115 correspond to unrounded values from 109.5 up to 115.5.

$P(110 \leqslant X \leqslant 115)$ *corresponds to* $P(109.5 \leqslant S < 115.5)$.

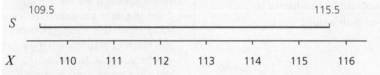

$S \sim N(100, 15^2)$

Using a calculator for

$P(109.5 \leqslant S < 115.5)$

```
1 : Normal PD
2 : Normal CD
3 : Inverse Normal
4 : Binomial PD
```

> 109.5 is the lowest number which rounds to 110. At the upper end of the range, numbers up to 115.5 will round to 115.

> Select Normal cumulative probabilities.

```
Normal    C.D
Data      : Variable
Lower     : 109.5
Upper     : 115.5
σ         : 15
μ         : 100
Save Res  : None
```

```
Normal  C.D
p       = 0.11253402
z : Low = 0.63333333
z : Up  = 1.03333333
```

Enter the mean, standard deviation and lower and upper limits.

The probability of a score between 110 and 115 (inclusive) is 0.113 (to 3 d.p.).

Examples of continuity corrections

Discrete variable, X	$X > 100$	$X \geqslant 100$	$X < 105$	$X \leqslant 105$
Approximating Normal, S	$S \geqslant 100.5$	$S \geqslant 99.5$	$S < 104.5$	$S < 105.5$

Common mistake: For the Normal distribution, it does not make a difference whether the inequality sign includes equals but it does make a difference for a discrete distribution.

Normal approximation to binomial distribution

A binomial distribution with p close to ½ is fairly symmetrical; for larger values of n, the binomial distribution is fairly symmetrical for values of p further from ½. The binomial distribution can be approximated by the Normal distribution; in the past this was used to make calculation of cumulative binomial probabilities easier. Modern software makes this approximation less useful but it is still sometimes used as people are often very accustomed to working with the Normal distribution.

Test yourself

TESTED

1 $X \sim N(5, 16)$. This means that X has a Normal distribution. Which one of the following statements is true?

   A   The mean is 4 and the standard deviation is 5

   B   The mean is 5 and the standard deviation is 4

   C   The mean is 5 and the standard deviation is 16

   D   The mean is 16 and the standard deviation is 5

   E   The mean is 16 and the standard deviation is 2.24 (3 s.f.)

2 Adult men in the UK have a mean height of 175 cm. Adult women in the UK have a mean height of 161.6 cm. The standard deviation for each population is 7.5 cm. You can assume that the heights are Normally distributed. An adult man and an adult woman are chosen at random. What is the probability that the man is shorter than 170 cm but the woman is taller than 170 cm? (Answers are given to 3 d.p.)

A   0.033   B   0.205   C   0.384   D   0.404   E   0.683

3 A machine makes bolts which should be 5 cm long. 15% of the bolts produced are shorter than 4.75 cm. The manager thinks that the machine is faulty and that the mean is different from 5 cm. The lengths of bolts produced are Normally distributed with standard deviation 0.07 cm. What is the mean? (Answers are in cm to 3 d.p.)

A   4.677   B   4.711   C   4.722   D   4.755   E   4.823

4 Three of the following statements are false and one is true. Find the one that is true.

A   The Normal distribution cannot be used as an approximation to a discrete variable.
B   Any symmetrical distribution can be approximated by a Normal distribution.
C   If a distribution is symmetrical and unimodal then it has an exact Normal distribution.
D   The histogram below shows the lengths of a random sample of 15 pea pods. The lengths of the whole population of pea pods could be Normally distributed.

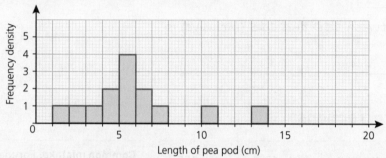

5 Marks awarded in a test can only be integers. The distribution of marks is approximately Normal with mean 45.4 and standard deviation 16. A mark of 70 or over is awarded a distinction. What proportion of candidates for the test is awarded a distinction? (Answers are given to 1 d.p.)

A   5.8%   B   6.2%   C   6.6%   D   11.9%   E   46.2%

Full worked solutions online

CHECKED ANSWERS

Exam-style question

A random sample of 180 students is asked to complete a puzzle. The times taken are shown in the table below.

Time, x minutes	$0 < x \leq 2$	$2 < x \leq 3$	$3 < x \leq 4$	$4 < x \leq 5$	$5 < x \leq 8$
Frequency	2	52	82	27	17

i   For these data, estimate
   ● the mean
   ● the standard deviation.
ii The times taken to complete the puzzle are modelled by a Normal distribution, with mean and standard deviation the values you found in part **i**. What is the probability of a randomly chosen student taking between 4 and 5 minutes, according to the model?
iii Do you think that the Normal distribution is a good model for these data? Give a reason for your answer.

Short answers on page 223

Full worked solutions online

CHECKED ANSWERS

Chapter 7 Extending hypothesis testing

About this topic

The ideas of hypothesis testing, which you have already used for situations which can be modelled by a binomial distribution, can be used in other situations as long as there is an underlying probability model. Using a sample mean to test the value of a population mean using a Normal probability model has wide applications because the Normal distribution is a commonly used distribution in all kinds of situations.

Scatter diagrams show relationships between two variables; considering how two variables are related to each other is important in statistical modelling to make predictions. Correlation coefficients are used to measure the strength of relationship in a scatter diagram; hypothesis testing can be used to judge whether the correlation from a sample is high enough to convince us that there is a relationship in the population.

Before you start, remember ...

- If a random variable, X, has a Normal distribution with mean μ and standard deviation σ, this can be written as $X \sim N(\mu, \sigma^2)$.
- How to work with probabilities.
- Correlation from GCSE.
- Ranked data from chapter 2.
- Straight-line graphs.

Hypothesis testing using the Normal distribution

REVISED

> **Key facts**
>
> 1 For samples of size n from $N(\mu, \sigma)$ the sample mean is Normally distributed with mean μ and variance $\dfrac{\sigma^2}{n}$, that is $\bar{X} \sim N\left(\mu, \dfrac{\sigma^2}{n}\right)$.
>
> 2 If the population variance is not known then the sample variance, s^2, can be used as an estimate, as long as n is large. In that case, $\bar{X} \sim N\left(\mu, \dfrac{s^2}{n}\right)$.
>
> 3 The null hypothesis for a test using the Normal distribution is that the population mean, μ, takes a particular value, e.g. $H_0 : \mu = 7.4$.
>   - For a one-tail test, the alternative hypothesis corresponding to the null hypothesis above would be EITHER $H_1 : \mu > 7.4$ OR $H_1 : \mu < 7.4$.
>   - For a two-tail test, the alternative hypothesis corresponding to the null hypothesis above would be $H_1 : \mu \neq 7.4$.
>
> 4 To conduct the test, a random sample of size n is taken and the sample mean, $\bar{x}$, is found. This can be used as the test statistic or it can be standardised and then used as a test statistic.
>
> 5 If the population has a Normal distribution with standard deviation σ, the test statistic for the null hypothesis given above is $z = \dfrac{\bar{x} - 7.4}{\dfrac{\sigma}{\sqrt{n}}}$. If the null hypothesis is true, this has a standard Normal distribution, with mean 0 and standard deviation 1.

What is the sampling distribution of the sample mean?

If a random sample of a given size, say 5, is taken from a Normal distribution with mean 30 and variance 9, then you would expect the sample mean to be somewhere near 30. Sometimes it might be a bit less than 30 and sometimes it might be a bit more than 30. The sample means for a large number of such samples are shown in the histogram.

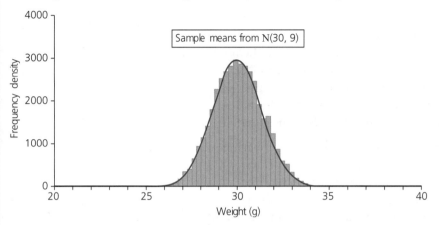

For samples of size n from $N(\mu, \sigma^2)$, the sample mean is Normally distributed with mean μ and variance $\dfrac{\sigma^2}{n}$, i.e. $\bar{X} \sim N\left(\mu, \dfrac{\sigma^2}{n}\right)$. This is called the **sampling distribution** of the sample means.

It looks as though the sample means are Normally distributed and that is, in fact, the case. The distribution of the sample means has mean 30 and variance $\dfrac{9}{5} = 1.8$.

Worked example

Example 1

Gherkins have a mean weight of 30 g. Assume that the weight of a randomly chosen gherkin is Normally distributed with standard deviation 3 g. A jar of gherkins contains 15 gherkins. What is the distribution of the mean of the weights of the gherkins in a jar? Assume that the gherkins in a jar are a random sample.

Solution

Let X be the weight, in grams, of a randomly chosen gherkin.

$X \sim N(30, 3^2)$

Samples are of size 15 so $n = 15$.

The distribution of the sample mean, $\bar{X}$, is $\bar{X} \sim N\left(30, \dfrac{3^2}{15}\right)$ i.e. N(30, 0.6).

Summarise the information in the question. Remember the variance is the square of the standard deviation.

Upper case $\bar{X}$ stands for the general sample mean, lower case $\bar{x}$ would stand for a particular value.

Information from samples

Samples are often taken to give information about likely values of population parameters, such as the population mean.

	True population value (this may be unknown)	Estimated value from the sample
Mean	μ	$\bar{x} = \dfrac{\sum x}{n}$
Variance for individual values	σ^2	s^2
Variance for mean of samples of size n	$\dfrac{\sigma^2}{n}$	$\dfrac{s^2}{n}$

Hint: Notice that for larger sample sizes from the same population, the variance of the sample mean is smaller. This fits in with the intuition that larger samples should give a more reliable estimate of the population mean.

Hint: If you are taking a sample from a population to find out about the population mean, then it may be the case that you do not know the population standard deviation. If the sample size is large enough then the sample standard deviation can be used as an estimate of the population standard deviation. Samples with at least 50 data values are usually large enough but the estimate can be reasonable for 30, or more.

Worked example

Example 2

A particular species of fish is being studied by fishery experts. They catch a random sample of 150 fish from this species and measure the length, x cm, of each. Here is a summary of the data:

$\bar{x} = 16.039$ (3 d.p.), variance = 19.756

The population mean for this species used to be 16.50 cm. If the population mean is still 16.50 cm, what is the probability of getting a sample mean less than 16.04 cm when taking a sample of 150 fish?

Solution

If the population mean is 16.50 cm then the sample mean for samples of size 150 has a Normal distribution with mean 16.50 and variance $\dfrac{19.756}{150}$.

Notice that you divide by $n = 150$ here.

This can be written $\bar{X} \sim N\left(16.50, \dfrac{19.756}{150}\right)$

The required probability is $P(\bar{X} < 16.04)$
Using cumulative Normal probabilities on your calculator:

Normal	C.D
Data	: Variable
Lower	: −1E+05
Upper	: 16.04
σ	: 0.362914
μ	: 16.5
Save Res	: None

Normal C.D	
P	= 0.102485

The probability of a sample mean less than 16.04 cm is 0.102 (3 s.f.)

Hint: For the standard deviation, you need $\sqrt{\dfrac{19.756}{150}}$. You can type this in as a calculation.

Common mistake: The variance of the sample mean is the population variance divided by n, $\dfrac{\sigma^2}{n}$. The standard deviation of the sample mean is the square root of this variance, $\sqrt{\dfrac{\sigma^2}{n}} = \dfrac{\sigma}{\sqrt{n}}$; it is not the population standard deviation divided by n.

1-tail test using a critical region

The distribution of the sample mean is needed when doing a Normal hypothesis test.

Worked example

Example 3

A factory makes a variety of battery which it claims has a mean lifetime of 7.4 hours of continuous use. The factory has changed one of its suppliers. The quality control inspector is concerned that the mean life for batteries could have been reduced. The inspector takes a sample of 5 batteries and tests how long they last for. The times, in hours, are:

    7.02   6.43   7.35   7.97   7.92

i   State suitable null and alternative hypotheses for a test.

ii Carry out the test at the 5% level of significance.

Assume that battery life is Normally distributed with standard deviation 0.6 hours.

Solution

i H_0: μ = 7.4 where μ is the mean battery life in hours for the population of batteries made by the factory.
    H_1: μ < 7.4

> The inspector suspects that the mean battery life could be lower now, so the alternative hypothesis is that μ is less than 7.4.

> **Hint:** The null hypothesis for this type of test is always that μ = a particular value. You should say that μ stands for the population mean; there is often a mark for doing so but do remember to say what the population is in the context of the question. Just saying 'μ is the population mean' with no further explanation is not enough.

ii

What you need to write down	Commentary	What to do on your calculator
If H_0 is true, $\bar{X} \sim N\left(7.4, \dfrac{0.6^2}{5}\right)$	If the null hypothesis is true, write down the distribution the sample mean comes from. Remember, for a sample of size n from $N(\mu, \sigma^2)$ the sample mean, $\bar{X}$, has a Normal distribution with mean μ and variance $\dfrac{\sigma^2}{n}$.	

> **Hint:** You could use the distribution of $\bar{X}$ to find the critical value for the lower 5% tail for the sample mean using inverse Normal probability on your calculator but if you standardise (as below) the 5% critical value will always be the same and this will help you spot any errors.

$\bar{x}$ = 7.338	Work out the sample mean.	Enter data values into a calculator and get the sample mean. $\begin{array}{ll} \bar{x} & = 7.338 \\ \Sigma x & = 36.69 \\ \Sigma x^2 & = 270.8951 \\ \sigma^2 x & = 0.332776 \\ \sigma x & = 0.5768674024 \\ s^2 x & = 0.41597 \end{array}$

What you need to write down	Commentary	What to do on your calculator
$z = \dfrac{\bar{X} - \mu}{\sqrt{\dfrac{\sigma^2}{n}}}$ $z = \dfrac{7.338 - 7.4}{\sqrt{\dfrac{0.6^2}{5}}} = -0.231\ldots$	Calculate the test statistic by standardising. (Subtract the mean and divide by the standard deviation). $Z = \dfrac{\bar{X} - \mu}{\sqrt{\dfrac{\sigma^2}{n}}}$ has a standard Normal distribution, N(0, 1).	
One-tail test	Is it a one or two-tail test? You will be able to tell from the form of the alternative hypothesis.	
  5%	Sketch the standard Normal curve with the tail you want shaded.	Use inverse Normal probability with 1 for standard deviation and 0 for mean. Inverse Normal Data : Variable Tail : Left Area : 0.05 σ : 1 μ : 0 Inverse Normal xInv $=-1.6448536$
Critical value = −1.645	Write down the critical value; you can get it from a table instead of using your calculator.	
−0.231 > −1.645 so not significant. There is insufficient evidence to reject H_0.	Make a decision by comparing the test statistic to the critical region.	
At the 5% significance level, the evidence does not support the suggestion that the mean battery life is reduced.	State the decision in context.	

Hint: There is often a mark given at the end of the question for stating the decision in context so don't just stop with 'do not reject H_0' or 'reject H_0'. Remember that you cannot say definitely whether the null hypothesis is true or false. Phrases like 'at the 5% significance level, the evidence does not support...' or 'there is sufficient evidence at the 5% level that...' are useful.

Using a *p*-value instead of a critical region

Hint: If you are finding a *p*-value, it will save time if you do not standardise but you should get the same probability whether you standardise or not.

For Example 3, above, if H_0 is true, $\overline{X} \sim N\left(7.4, \frac{0.6^2}{5}\right)$. The observed value of the sample mean was 7.338 and smaller values would cast even more doubt on the null hypothesis, so work out $P(\overline{X} < 7.338)$.

Use cumulative Normal probability on a calculator, entering the standard deviation as $\frac{0.6}{\sqrt{5}}$.

```
Normal     C.D
Data        : Variable
Lower       : −1E+05
Upper       : 7.338
σ           : 0.26832815
μ           : 7.4
Save Res    : None
```

```
Normal  C.D
p       = 0.40863395
z : Low = −3.727E+05
z : Up  = −0.2310603
```

This probability (0.4086 in this example) is called the **p-value**. The null hypothesis would be rejected if it was too small. The *p*-value is compared with the significance level to decide whether it is too small.

0.4086 is not a small probability; it is much bigger than 0.05 (the significance level). There is insufficient evidence to reject H_0. It is reasonable to conclude that the mean battery life is not reduced.

Hint: Remember the *p*-value is the probability of getting an outcome at least as extreme as that observed, if the null hypothesis is true.

2-tail tests using critical values

Worked example

Example 4

The mean wingspan of a particular species of butterfly is 6.2 cm. The standard deviation is 0.3 cm. A zoologist suspects that the mean wingspan for this species in a particular region has changed due to environmental factors. A sample of 10 butterflies is measured; the mean wingspan for the sample is 6.38 cm. Test at the 1% level of significance whether this provides evidence of a change in mean wingspan. Assume that the standard deviation is still 0.3 cm and that wingspans are Normally distributed. State your hypotheses and conclusions carefully.

Solution

$H_o : \mu = 6.2$

$H_1 : \mu \neq 6.2$ where μ is the mean wingspan in cm for the population of butterflies in the region.

Let X be the wingspan of a randomly chosen butterfly in cm.

If the null hypothesis is true, $X \sim N\left(6.2, 0.3^2\right)$

The distribution of the sample means is $\bar{X} \sim N\left(6.2, \dfrac{0.3^2}{10}\right)$.

The test statistic is $z = \dfrac{\bar{x} - \mu}{\sqrt{\dfrac{\sigma^2}{n}}} = \dfrac{6.38 - 6.2}{\sqrt{\dfrac{0.3^2}{10}}} = 1.8973....$

The zoologist is looking for evidence of a change in either direction, so it is a two-tail test.

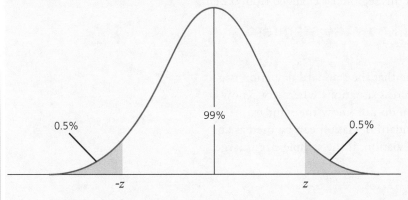

The critical values are $z = \pm 2.576$ and the critical region is $\{z : z < -2.576 \cup z > 2.576\}$

The test statistic, $1.8973...$, is not in the critical region so there is insufficient evidence to reject H_o

At the 1% significance level, there is insufficient evidence that the mean wingspan for this species of butterfly in this region has changed from 6.2 cm.

> The zoologist is looking for evidence of a change.

> The sample size, n, is 10.

> **Hint:** You were told that the sample mean was 6.38 cm so $\bar{x} = 6.38$. There were 10 butterflies in the sample so $n = 10$.

> **Common mistake:** Don't get confused between the sample mean, $\bar{x}$, and the population mean, μ.

> The significance level is 1% and it is a two-tail test so 0.5% for each tail.

> **Hint:** You can get the critical values using inverse Normal probability on a calculator or from tables. Notice that the two critical values are symmetrical about zero for the standard Normal.

p-value for a 2-tail test

For Example 4, above, if H_0 is true, $\bar{X} \sim N\left(6.2, \dfrac{0.3^2}{10}\right)$. The observed value of the sample mean was 6.38; this is bigger than the value of the population mean in the null hypothesis so larger values would cast even more doubt on the null hypothesis so work out $P(\bar{X} > 6.38)$.

Use cumulative Normal probability on a calculator, entering the standard deviation as $\dfrac{0.3}{\sqrt{10}}$.

Normal	C.D
Data	: Variable
Lower	: 6.38
Upper	: 1E+06
σ	: 0.09486832
μ	: 6.2

```
Normal  C.D
p      =0.02888978
z : Low=1.8973666
z : Up =1.0541E+07
```

$P(\overline{X} > 6.38) \approx 0.0289$

> **Common mistake:** This is a two-tail test so this probability is not the p-value;
> it could be compared to half the significance level to make a decision
> but, because the Normal distribution is symmetrical, you can double this
> probability to get the p-value and then compare to the significance level.

p-value $= 2 \times 0.0289 = 0.0578 > 0.01$

At the 1% significance level, there is insufficient evidence that the mean
wingspan for this species of butterfly in this region has changed from 6.2 cm.

What happens if you do not know the standard deviation?

In Example 4, you were told to assume that the standard deviation was
unchanged (0.3 cm). You may come across situations where you know
that the distribution is Normal but you do not know the standard
deviation. In that case, the sample standard deviation can be used as an
estimate of the population standard deviation, if the sample size is large
enough (50 is usually large enough).

Test yourself

TESTED

1 Random samples of size 5 are taken from N(12, 2.5). What is the distribution of the mean of such
   samples?

   A   N(2.4, 0.5)   B   N(12, 0.5)   C   N(12, 0.1)   D   N(12, 2.5)   E   N(60, 12.5)

2 A man has found that the time it takes him to walk to work is Normally distributed with mean 15.5
   minutes. After going on a fitness course, he thinks that he may be walking faster on average now. Which
   are suitable null and alternative hypotheses to use for a test?

   A $H_0 : \mu = 15.5$ B $H_0 : \mu = 15.5$ C $H_0 : \mu = 15.5$ D $H_0 : \mu < 15.5$ E $H_0 : \mu > 15.5$
       $H_1 : \mu < 15.5$ $H_1 : \mu > 15.5$ $H_1 : \mu \neq 15.5$ $H_1 : \mu > 15.5$ $H_1 : \mu < 15.5$

3 Which is the best description of the meaning of μ in the hypotheses in Question 2?

   A   The critical value.

   B   His new mean walking time, in minutes, after the fitness course.

   C   The mean of a random sample of his walking times, in minutes, after the fitness course.

   D   The population mean.

   E   The test statistic.

This information is for Questions 4 and 5.

When a coffee machine is working properly, the volume of coffee dispensed is Normally distributed with mean 175 ml and standard deviation 6.4 ml. After the machine is serviced, a sample of 10 cups of coffee is taken. The volumes of coffee, in ml, are as follows:

| 178 | 172 | 182 | 188 | 188 | 191 | 166 | 179 | 187 | 173 |

Test at the 10% significance level whether the mean amount of coffee dispensed differs from 175 ml after the service. Assume that the volumes of coffee are Normally distributed with standard deviation 6.4 ml.

4 Which of the following is the standardised test statistic for the test described above? (Rounded answers are given to 4 decimal places)

   A $z = 0.8438$ B $z = 1.3184$ C $z = 2.6682$ D $z = 8.4375$

Make sure you have got the right answer to Question 4 before doing Question 5

5 Which one of the following statements is true for the test described above?

   A   The test statistic is in the critical region.

   B   If another sample of size 10 was taken, the critical region could have been different.

   C   The test statistic for this sample would be different if the significance level was 2%.

   D   The test statistic will be the same for all samples of size 10.

   E   The null hypothesis must be false.

Full worked solutions online

CHECKED ANSWERS

Exam-style question

A national survey finds that the heights of men are Normally distributed with mean 176 cm and standard deviation 6.7 cm.

A historian is studying army records and he suspects that the soldiers he is studying were shorter, on average.

A random sample from the records gives the following heights of soldiers (all in cm):

| 165 | 158 | 169 | 176 | 175 | 179 | 172 |

i Write down suitable null and alternative hypotheses to test the historian's suspicion.

ii Assume that the standard deviation of the heights of soldiers is 6.7 cm. Carry out a suitable hypothesis test, at the 5% level of significance, to test the historian's suspicion. You should state your conclusions clearly.

Short answers on page 223

Full worked solutions online

CHECKED ANSWERS

Bivariate data

Key facts

1 Bivariate data are in pairs so they involve two variables, e.g. age and height of a sample of children. Such data can be plotted on a scatter diagram.

2 The **independent** variable goes on the x-axis of a scatter diagram and the **dependent** variable goes on the y-axis.

3 An outlier in a scatter diagram is a point that does not fit the general pattern of the rest of the data.

4 Correlation means a linear relationship. Pearson's product moment correlation coefficient (often just called the correlation coefficient) measures the strength of the linear relationship. It can take values between -1 and $+1$ (inclusive).

5 A correlation coefficient of $+1$ means the points all lie in a straight line with positive gradient. A correlation coefficient of -1 means the points all lie in a straight line with negative gradient.

6 A correlation coefficient of 0 means there is no correlation, i.e. no linear relationship.

7 For a test using the correlation coefficient, the null hypothesis, H_0, is that $\rho = 0$ where ρ is the population correlation coefficient. The null hypothesis is that there is no correlation in the population.

8 The alternative hypothesis, H_1, will be one of the following:
- $\rho > 0$ (one-tail test, testing for positive correlation)
- $\rho < 0$ (one-tail test, testing for negative correlation)
- $\rho \neq 0$ (two-tail test, testing for some correlation).

9 The critical value for the correlation coefficient tells you how close to 1 or -1 the sample correlation coefficient has to be for you to reject the null hypothesis that $\rho = 0$.

10 Even if there is strong correlation, it does not mean that change in one of the variables **causes** change in the other variable.

11 A scatter diagram might reveal a relationship between the variables which is not linear; association means any relationship between the variables.

12 If a relationship is not linear, then there is often correlation between the ranks of the data.

13 A data set which consists of values of a variable together with the times at which they occurred is a time series; the time is plotted on the horizontal axis. If the dependent variable varies continuously with time, the points may be joined (dot to dot) to show the trend.

Patterns in scatter diagrams

Positive correlation

Negative correlation

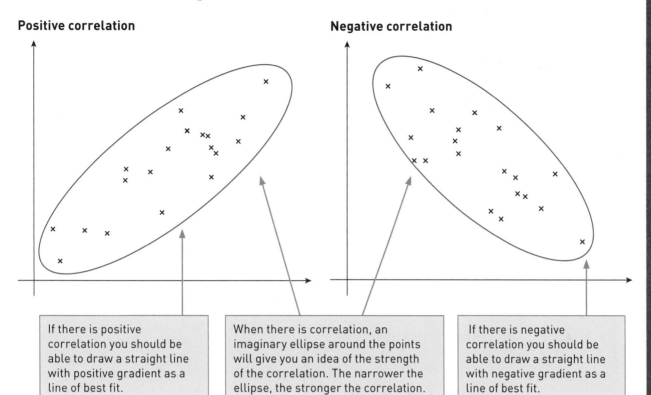

If there is positive correlation you should be able to draw a straight line with positive gradient as a line of best fit.

When there is correlation, an imaginary ellipse around the points will give you an idea of the strength of the correlation. The narrower the ellipse, the stronger the correlation.

If there is negative correlation you should be able to draw a straight line with negative gradient as a line of best fit.

No correlation

Two sections

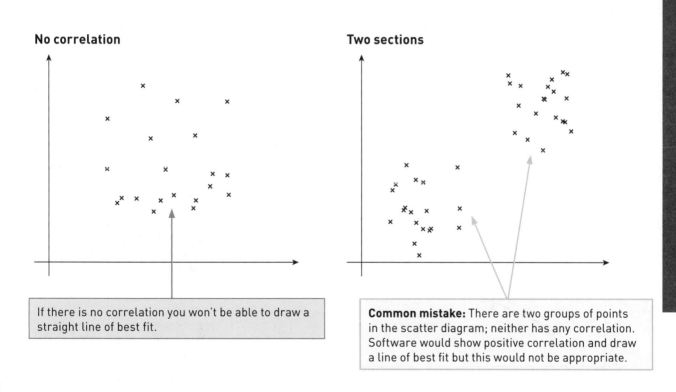

If there is no correlation you won't be able to draw a straight line of best fit.

Common mistake: There are two groups of points in the scatter diagram; neither has any correlation. Software would show positive correlation and draw a line of best fit but this would not be appropriate.

Measuring correlation

A correlation coefficient measures the strength of a linear relationship. The correlation coefficient is:

- +1 for a perfect straight line with positive gradient
- −1 for a perfect straight line with negative gradient
- 0 if there is no correlation.

Correlation coefficients can be calculated easily using calculators or software such as spreadsheets.

> You do not need to know how correlation is calculated but you need to be able to interpret given values of correlation coefficients.

Outliers

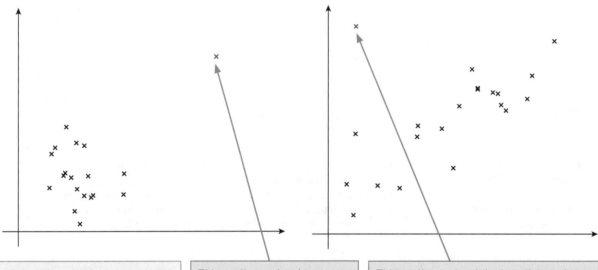

> **Hint:** Any outliers in the scatter diagram should be investigated; they may be errors or they may be genuine values.

> This outlier makes it appear as if there is correlation but there is no correlation for the other points.

> This outlier makes it look as if there is less correlation than there is for the rest of the data. The correlation coefficient with the outlier is 0.587 and without the outlier is 0.8135.

Using best fit models

The scatter diagram below shows a sample of car journeys completed by a salesman. The distance in miles and the time taken in minutes are shown. A line of best fit has been drawn using a spreadsheet.

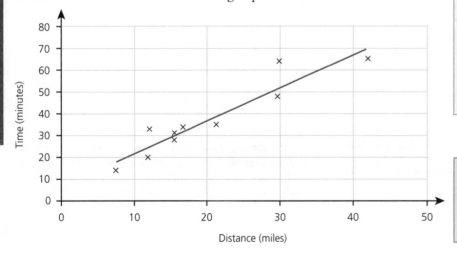

> **Hint:** The independent variable goes on the x-axis. In this case, the salesman is interested in how time taken depends on distance so distance is the **independent variable** and time taken is the **dependent variable**.

> When a line of best fit is calculated (for example using software or calculator functions) it is known as a *regression line*.

Interpolation and extrapolation

Interpolation is estimating for a data point within the range of points you already have. For the salesman data, using the regression line to estimate the time taken for a journey of 35 miles would be interpolation.

Extrapolation is using the regression line beyond the range of points you already have. For the salesman data, using the regression line to estimate the time taken for a journey of 100 miles would be extrapolation.

> Results from interpolation are usually reasonable estimates but results from extrapolation may be poor estimates, especially if they are a long way from the data.

Worked example

Example 1

The equation of the line of best fit for the salesman is $y = 1.5x + 6.8$, where x is the distance in miles and y is the time taken in minutes.

i   Give an interpretation of the gradient of the line.
ii What does 6.8 in the equation represent?
iii Estimate the time taken for a journey of:
   A 25 miles   **B** 200 miles.
iv Comment on the reliability of your answers.

> **Hint:** Remember that in the form $y = mx + c$, m represents the gradient. The gradient is the increase in y for a one unit increase in x.

Solution

i   The gradient of $y = 1.5x + 6.8$ is 1.5; it represents the time taken to travel each mile.

ii 6.8 minutes is the time for no miles so it represents the time taken to start the journey.

iii **A** $x = 25 \Rightarrow y = 1.5 \times 25 + 6.8 = 44.3$.
       44.3 minutes
    B $x = 200 \Rightarrow y = 1.5 \times 200 + 6.8 = 306.8$
       307 minutes (3 s.f.). This is just over 5 hours.

iv The answer for 25 miles is likely to be fairly reliable because 25 is in the range shown in the scatter diagram. The answer for 200 miles will be less reliable as it is further from the data; it assumes the straight-line relationship will continue but longer journeys may take less time for each mile as part of the journey could be on the motorway.

> **Hint:** If the journey had no distance at all then it would take no time but the 6.8 minutes is when the distance on any journey starts increasing. It is the time taken to get everything in the car and set off.

> **Hint:** When using a model, it isn't appropriate to give your answer to a lot of decimal places; in this case the time taken for a given journey would probably vary quite a bit. It is also more natural to state a long time like this in hours rather than in minutes.

> **Common mistake:** A line of best fit or other model is often used for making predictions so there may be some extrapolation; the further you extrapolate, the more cautious you should be about your answer.

Worked example

Example 2

The scatter diagram below shows the median age and the birth rate per 1000 in all the countries of the world for which data are available. A line of best fit has been drawn using a spreadsheet.

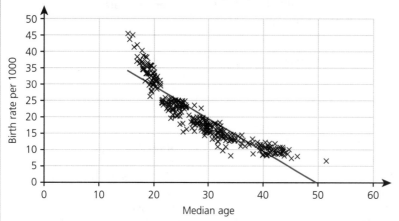

i   Describe the correlation in the scatter diagram.
ii Outline the limitations of the straight-line model.
iii Sketch the graph of a better model.

Hint: Although the straight-line model could be improved, all points do lie fairly close to the line and the line has a negative gradient so there is strong negative correlation.

Hint: Look at different sections of the data set where the straight-line model could be improved.

Hint: A curved model is better.

Common mistake: Do not continue the model beyond the range of the data; it might not be appropriate. In this case, there are no countries with higher or lower medians than those shown.

Solution

i   *There is strong negative correlation.*
ii *For countries with a high or low median age, the straight line suggest a lower birth rate than the actual birth rate but for countries with a medium median age, the line suggests a birth rate that is generally too high.*

iii
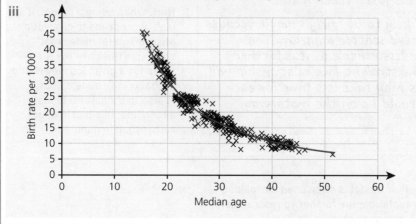

For relationships which are not straight line relationships, we say there is strong association if all the points lie near the curve, as in Example 2 above.

Hypothesis tests using correlation coefficients

Worked example

Example 3

Verna is making slippers for children. She measures the feet of a random sample of 10 girls. The results are shown in the following scatter diagram.

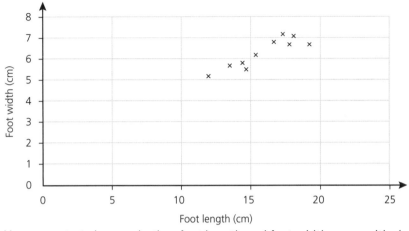

Verna wants to know whether foot length and foot width are positively correlated for girls.

i State suitable null and alternative hypotheses for a suitable statistical hypothesis test.

ii The correlation coefficient for the sample is 0.8958. The critical value for a one-tail test at the 5% level is 0.5494. Is there evidence of correlation in the population?

Solution

i $H_o : \rho = 0$

$H_1 : \rho > 0$ where ρ is the population correlation coefficient

ii 0.8958 > 0.5494 so reject the null hypothesis.
There is sufficient evidence, at the 5% significance level, of positive correlation between girls' foot length and width.

Hints:
- The null hypothesis is that there is no correlation in the population.
- Verna wants to know whether there is positive correlation in the population so the alternative hypothesis is that there is positive correlation in the population.
- Remember to say what ρ stands for.

Hints:
- The process of hypothesis testing, using either a critical value or a p-value, is the same as for the other hypothesis tests you have done.
- Critical values are usually given as positive; if the size of the correlation coefficient is bigger than the critical value, this provides evidence for accepting the alternative hypothesis.

Common mistake: There is clearly positive correlation for the sample; that can be seen from the scatter diagram without even calculating the correlation coefficient. There often appears to be correlation in small samples even if there is very little correlation in the whole population, so the hypothesis test helps you judge whether the correlation in the sample is high enough to convince you that there is really correlation in the population.

Interpreting a correlation coefficient

If there is evidence of correlation it does not mean that change in one of the variables causes change in the other variable. For example, there is a correlation between children's shoe size and reading ability. This is because older children tend to have bigger feet and they also tend to be better at reading; it is not because having bigger feet makes you a better reader.

If there is no correlation, this does not mean there is no relationship between the variables. The relationship might not be a straight-line (linear) relationship.

Time series

A time series consists of the values of a variable taken at different times. The values are plotted with the time on the horizontal axis. To show the trend over time, the points can be joined (dot to dot); a broken line should be used if the data values only exist at the points given. Several time series can be plotted on the same graph to allow comparison.

Worked example

Example 4

The graph below shows the height of a river in Texas from 1 August to 29 August 2017.

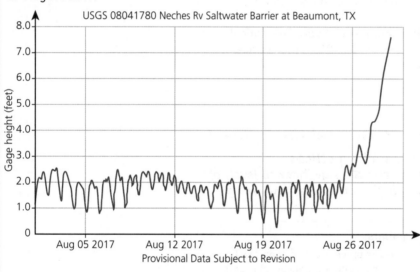

Make two comments about the trends shown in the height of the river.

Solution

The height of the river varies over the course of a day.

From 26 August onwards, the river is getting gradually higher; it becomes much higher than usual.

> **Hint:** The question asks you to make two comments. There are more than two comments which could be made in this context but, in an examination, you should give the number of comments asked for.

Test yourself

1 Fred exercises for 30 minutes then observes his pulse rate after stopping the exercise. The data are shown in the table and scatter diagram below.

Time after exercise (min), x	Pulse rate, y
½	133
1	127
1½	98
2	67
2½	68

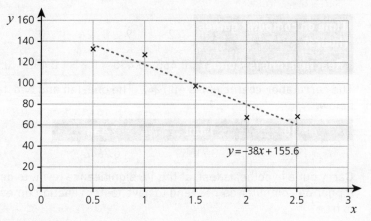

$y = -38x + 155.6$

Three of the following statements are false and one is true. Find the one that is true.

A It would be reasonable to use the equation of the regression line to find the pulse rate 4 minutes after exercise as this is not far from the data in the table.

B The equation of the regression line which you have found will apply to the pulse rate of all men who exercise for 30 minutes then stop.

C Time after exercise is the dependent variable and pulse rate is the independent variable.

D 79.6 is the model's prediction of the pulse rate 2 minutes after Fred stops exercise.

2 A researcher is testing whether there is a connection between mathematics ability and ability to solve Sudoku puzzles quickly. A sample of 11 students sit a mathematics test and solve a Sudoku. The results are shown in the scatter diagram below.
If the point representing the student who took 36 minutes to solve the Sudoku was removed, what effect would this have on the correlation coefficient?

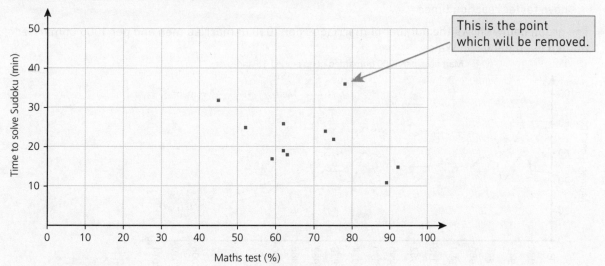

This is the point which will be removed.

A The correlation coefficient for the 10 remaining students is nearer to –1 than it was for all 11 students.

B The correlation coefficient for the 10 remaining students is nearer to 0 than it was for all 11 students.

C The correlation coefficient for the 10 remaining students is nearer to 1 than it was for all 11 students.

D The only way to decide which of A, B or C is true is to work out the correlation coefficient for all 11 students and then for the 10 remaining students.

3 A student believes that people who play more computer games tend to have faster reaction times. To test the theory, he asks a random sample of 10 people how long each spent playing computer games in the last week. He also carries out a reaction time test with each one. The results are in the table below.

Time on computer games (hours)	2	9	24	0	1	7	10	20	14	7
Reaction time (seconds)	0.465	0.435	0.31	0.485	0.49	0.355	0.21	0.2	0.26	0.285

The correlation coefficient is −0.7242. The one-tail and two-tail critical values are shown in the table below.

n = 10	1-tail	2-tail
1%	0.7155	0.7646

Carry out a hypothesis test, at the 1% significance level, to determine whether people who spend longer on computer games tend to have faster reaction times. Which one of the following statements is true?

A   The sample correlation coefficient is negative so there is no evidence that people who spend longer on computer games tend to have faster reaction times.

B   This is a one-tail test. The size of the sample correlation coefficient is bigger than the critical value so there is sufficient evidence that people who spend longer on computer games tend to have faster reaction times.

C   This is a one-tail test. The sample correlation coefficient is smaller than the critical value so there is insufficient evidence that people who spend longer on computer games tend to have faster reaction times.

D   This is a two-tail test. The sample correlation coefficient is bigger than the critical value so there is sufficient evidence that people who spend longer on computer games tend to have faster reaction times.

E   This is a two-tail test. The size of the sample correlation coefficient is smaller than the critical value so there is insufficient evidence that people who spend longer on computer games tend to have faster reaction times

4 The graph below shows the number of marriages per 1000 unmarried men and per 1000 unmarried women in England and Wales.

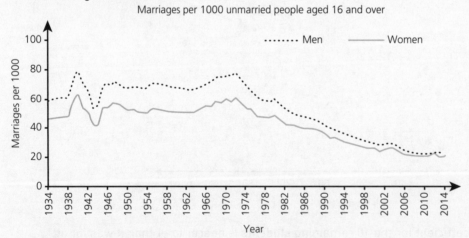

Marriages per 1000 unmarried people aged 16 and over

Four of the following statements are true and one is false. Find the one that is false.

A   The graph shows that in 2014 about one in every 50 unmarried adults got married.

B   The graph shows that more men than women get married.

C   It is possible to infer that, for the time period shown by the graph, there were more unmarried women than unmarried men in England and Wales.

D If a scatter diagram was plotted with marriages per 1000 men on the horizontal axis and marriages per 1000 women on the vertical axis then there would be very strong positive correlation.

E Marriages per 1000 unmarried people shows the trend in marriages better than total marriages in the population as it eliminates the effect of population change on the statistics.

Full worked solutions online

CHECKED ANSWERS

Exam-style question

Eight history students are chosen at random to sit two examination papers. Their marks are shown in the scatter diagram.

i Describe the correlation in the scatter diagram.

ii One of the students did not notice the back page of questions on Paper 2.

A Which point in the scatter diagram represents the marks of this student?

B Should the data from this student be used when testing for correlation in the population? Justify your answer.

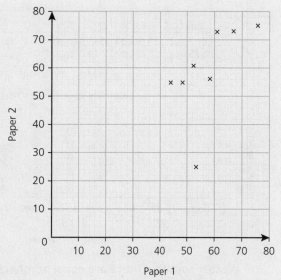

The examiners say that there should be a positive correlation between the marks on both papers.

iii The correlation coefficient for the data from the seven students who completed all questions is 0.8687. The p-value for a 1-tail test is 0.0056. Test, at the 5% level, whether the examiners' claim is true. You should state your hypotheses and conclusions clearly.

iv The examinations were tried out on a different random sample of students. Both the correlation coefficient and the p-value were lower. What can you say about the number of students in the second random sample? Give a reason for your answer.

Short answers on page 223

Full worked solutions online

CHECKED ANSWERS

Review questions (Statistics)

1 Teachers in England are classified as leadership teachers if they are heads, deputy heads, assistant heads or advisory teachers. Other teachers are classified as classroom teachers.

The charts below show the number of teachers in state secondary schools in England in 2016.

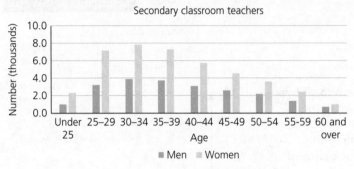

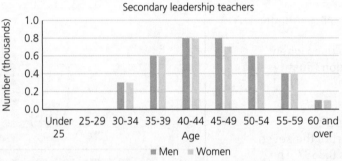

i Wesley says that there are equal numbers of men and women in leadership positions so women teachers and men teachers are equally likely to go into leadership in secondary schools. Comment on Wesley's statement.

ii An advertising campaign is being planned to increase the number of teachers at secondary level. Suggest what kind of people it should be targeted at. Justify your answer.

2 A sample of students are told to close their eyes; they are told to put up a hand when they think a minute has gone by. The times it takes them to put up their hands are summarised in the following table and histogram.

Time (x seconds)	Number of students
$20 \leqslant x < 30$	2
$30 \leqslant x < 40$	4
$40 \leqslant x < 45$	3
$45 \leqslant x < 50$	6
$50 \leqslant x < 60$	7
$60 \leqslant x < 70$	6
$70 \leqslant x \leqslant 100$	2

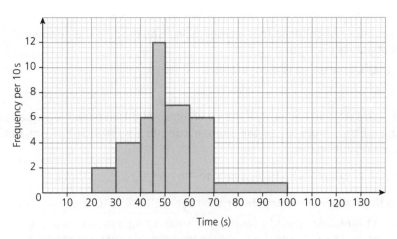

i Calculate an estimate of the median of the data.

ii Calculate an estimate of the mean of the data.

iii Joe suggests that a Normal distribution with a mean of 60 would be a good model for this situation. Comment on Joe's suggestion.

3 A random sample of word processed pages produced by a new secretary is being proof read. The random variable X denotes the number of errors per page.

The results are summarised as follows.

$$n=50, \sum x=53, \sum x^2 =103$$

Calculate:

i the mean

ii the standard deviation.

4 When George instructs his computer to copy a folder, it gives an estimate of how long it will take. George wants to see how accurate the estimates are. He measures the actual times taken for a random sample of folders to copy. The results are shown in the scatter diagram.

i Describe the correlation in the scatter diagram.

ii The correlation coefficient is 0.7638. George suspects that removing the two points with the largest estimated times will increase the correlation coefficient. Is he right? Justify your answer.

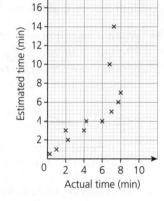

5 A and B are events with $P(A') = 0.3$, $P(B) = 0.5$, and $P(A \cup B) = 0.8$.

i Find $P(A)$.

ii Find $P(A \cap B)$.

iii Are A and B mutually exclusive? Explain how you know.

6 A magazine conducts a survey of a random sample of its readers to find out if they would be interested in subscribing to an internet version of the magazine. The results are summarised in the table below which shows the number of people in each category.

	Age				Row totals
	under 25	26–40	41–60	over 60	
Would subscribe to internet version?	11	60	27	38	136
Would not subscribe to internet version?	45	113	56	50	264
Column totals	56	173	83	88	400

Assume that the sample is representative of all the readers.

i What is the probability that a randomly chosen reader would subscribe to the internet version?

ii What is the probability that a randomly chosen reader over 60 would subscribe to the internet version?

7 A statistician is investigating the number of children that women have. She models the number of children per woman, X, by the probability distribution

$P(X=r)=k(5-r)(4+r)$ for $r=0$, 1, 2, 3, 4, where k is a constant.

  i   Find the value of k.

  ii Another statistician says that some women have more than 4 children and that possible values of r should include 5 and 6. Would it be possible to use the same model, $P(X=r)=k(5-r)(4+r)$, but with a different value of k? Explain your answer.

8 An airline finds that, on average, 5% of people who book a seat fail to turn up for the flight.

  i   If 142 bookings are taken for a flight, how many people, on average, will fail to turn up?

142 bookings have been taken for a flight. You should assume that individual passengers are a random sample from the population and turn up for the flight independently of each other, with equal probability.

  ii The aeroplane has 136 seats. Find the probability that more than 136 people turn up for the flight.

9 A survey shows that the systolic blood pressure of women in the UK is Normally distributed with mean 121 mm Hg and standard deviation 16.2 mm Hg.

  i   Hypertension is defined as a systolic blood pressure over 140 mm Hg. What proportion of women in the UK suffer from hypertension?

  ii Ten women are chosen at random from the population. Find the probability that at least one of them has hypertension.

10 A new fruit picker starts work. 8% of the baskets of strawberries she picks take her longer than 15 minutes. 5% of them take less than 10 minutes.

Assuming that the time the new picker takes to pick a basket of strawberries is Normally distributed, find the mean and standard deviation of her picking times.

11 When a biscuit production line is working correctly it produces biscuits with a mean mass of 20 g and standard deviation 0.95 g. The masses of biscuits are Normally distributed. To check the process is working correctly, a random sample of 60 biscuits is taken and the mass, x g, of each biscuit is measured. The sample mean is 19.7 g.

  i   State suitable null and alternative hypotheses for a test to see whether the mean mass of biscuits produced has changed.

  ii Carry out the test, at the 5% significance level, stating your conclusions carefully. You should assume that the standard deviation is unchanged.

12 The following table shows the mean maximum temperature in April and the total April rainfall. The data are for a random sample of 12 years for the same place.

Temp (°C)	9.3	14.5	11.4	11.6	13.3	13.5	13.9	18.1	15.2	12.0	15.4	14.0
Rain (mm)	23.8	10.4	28.1	97.7	19.6	43.1	70.5	1.6	35.9	55.7	19.2	9.6

  i   A researcher thinks that warmer Aprils tend to have less rain. Write down suitable hypotheses for a test of this using a correlation coefficient.

  ii The correlation coefficient is -0.4125. The critical value for a 1-tail test using the correlation coefficient is 0.4973 at the 5% level of significance. What is the conclusion for the hypothesis test?

Short answers on pages 223–224
Full worked solutions online

CHECKED ANSWERS

MECHANICS

1 Drawing a displacement–time graph

A boy runs 100 m from A to B at a constant velocity of 5 m s⁻¹. He waits for 5 s at B and then runs back to A at 4 m s⁻¹.

i   Find the total time the boy takes from leaving A to returning there.

ii Draw a displacement–time graph for the boy.

(see page 127)

2 Interpreting the gradient of a position–time graph

Megan and her father leave home together and go to a football field. Megan runs on ahead and waits for her father when she gets there. Their journeys are shown on the distance–time graph in the figure below.

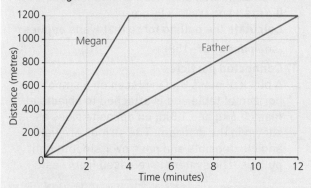

What is the difference in their speeds while they are travelling? Give your answer in m s⁻¹ correct to 2 s.f.

(see pages 126 and 128)

3 Interpreting the gradient of a speed–time graph

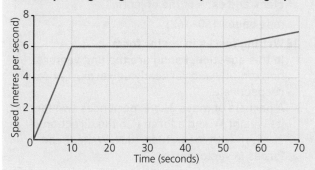

The graph in the figure above shows the speed of a runner in a race.

i   What was the magnitude of the runner's acceleration at each stage?

ii Calculate the distance travelled in total.

(see pages 129–131)

4 Using constant acceleration formulae to find acceleration and distance travelled

A car is initially travelling at 12 m s⁻¹. The driver takes her foot off the accelerator and the car comes to rest in 7.5 s. Assuming the acceleration of the car is constant, calculate:

i   the acceleration

ii the distance the car travels while coming to rest.

(see pages 134–136)

5 Finding the time taken and the final velocity for vertical motion under gravity

Atul drops an egg onto the floor from a shelf 1.2 m above the ground. Find:

i   the time it takes to reach the ground

ii the velocity of the egg when it hits the ground.

(see pages 138–140)

6 Understanding vector forms of displacement and velocity

In this question the x and y directions are East and North, respectively.

i   The initial position of a model boat in metres is given by $r=\begin{pmatrix} 5 \\ -12 \end{pmatrix}$. Find its distance and its bearing from the origin.

ii Initially the model boat is travelling at 4.5 m s⁻¹ on a bearing of 060°. Write down the initial velocity as a column vector.

Give your answers correct to 3 significant figures where necessary.

(see pages 164–170)

7 Using the constant acceleration formulae in 2 dimensions

A particle accelerates from $\mathbf{u}=9\mathbf{i}-5\mathbf{j}$ to $\mathbf{v}=-\mathbf{i}-\mathbf{j}$ in 4 seconds. Distances are in metres.

i   Find the acceleration of the particle.

ii Find the distance travelled in 4 seconds.

(see pages 164–170)

8 Using the position of a projectile at a given time

A ball is thrown from a point at ground level at 24 m s⁻¹ at an angle of 20° above the horizontal across level ground.

i   Write expressions for the horizontal distance x and the height y of the ball above its initial position at time t seconds.

ii A fence is 2 m high and 5 m away from the point of projection. Does the ball go over the fence?

(see pages 194–197)

9 Finding the greatest height and the range on level ground

A projectile is launched from ground level with an initial velocity $15\,\text{m}\,\text{s}^{-1}$ at an angle of $35°$ to the ground. The ground is a horizontal plane.

i   Find the maximum height reached by the projectile.

ii Find the distance travelled horizontally by the time that the projectile hits the ground.

(see pages 194–197)

10 Finding the cartesian equation of the trajectory of a projectile

A ball is thrown horizontally at $7\,\text{m}\,\text{s}^{-1}$ from a point which is $1.5\,\text{m}$ above the origin. Find the cartesian equation of the trajectory of the ball, using x and y for the horizontal and vertical displacements from the origin.

(see pages 194 and 197)

11 Using Newton's 3rd law

The diagram shows a pile of 3 blocks A, B and C in equilibrium on a rough horizontal table. Their masses are $2\,\text{kg}$, $3\,\text{kg}$ and $4\,\text{kg}$. The block C exerts a force N on the block B. State the magnitude and direction of N.

(see pages 148–150)

12 Identifying forces

A plank of wood rests on a fixed smooth cylinder with one end on rough horizontal ground as shown in the diagram. Show all the forces acting on the plank of wood.

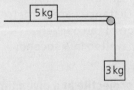

(see pages 148–151)

13 Using Newton's 1st law for forces in equilibrium

The diagram shows a block of mass $5\,\text{kg}$ on a rough horizontal table. It is attached by a light string to a block of mass $3\,\text{kg}$. The string passes over a light smooth pulley. The system is in equilibrium at rest.

Find the value of the frictional force acting on the larger block.

(see pages 149 and 151)

14 Using Newton's 1st law for forces in two directions in equilibrium

In this question, $\mathbf{i}$ is a horizontal unit vector and $\mathbf{j}$ is a unit vector vertically upwards. A particle of mass $4\,\text{kg}$ is in equilibrium under the action of three forces: its weight, $\mathbf{W}$ N, and the tensions in two strings that are attached to it, $\mathbf{T_1} = 5\mathbf{i} + y\mathbf{j}$ and $\mathbf{T_2} = x\mathbf{i} + 12\mathbf{j}$ in N.

Find the values of x and y.

(see pages 171–172)

15 Using Newton's 2nd law to find acceleration

The mass of a helicopter at the time it is launched is 2 tonnes. The rotary blades produce an upwards force of D N.

i   Find the greatest value of D for which lift-off will not be achieved.

ii The initial value of D is actually $23\,000$. Calculate the acceleration of the helicopter.

(see pages 157–158)

16 Using Newton's 2nd law to find a force

Iyana pushes a toy car of mass $0.7\,\text{kg}$ along a straight horizontal track with a horizontal force for $0.8\,\text{s}$. The resistance to motion is $5\,\text{N}$. The car accelerates from an initial speed of $0.4\,\text{m}\,\text{s}^{-1}$ and travels $80\,\text{cm}$.

Calculate the pulling force that Iyana applies.

(see pages 157–158)

17 Connected particles

A block of mass $3\,\text{kg}$ is placed on a smooth horizontal table and is attached to objects of mass $0.6\,\text{kg}$ and $1.8\,\text{kg}$ on opposite sides as shown in the diagram. The strings are light and inextensible and pass over smooth light pulleys. The system is released from rest.

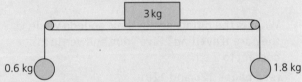

Write down the equations of motion for the block and each of the objects.

(see pages 160–162)

18 Writing forces in vector form

In this question, $\mathbf{i}$ and $\mathbf{j}$ are the unit vectors horizontally and vertically upwards, respectively.

A particle of mass $3\,\text{kg}$ experiences two forces, its weight $\mathbf{W}$ and a force $\mathbf{F}$ in the direction of $2\mathbf{i} + 1.5\mathbf{j}$ with magnitude 10 N. Write these forces in vector form.

(see pages 171–174)

19 Using the vector form of Newton's 2nd law to find acceleration

In this question all quantities are in S.I. units and the x and y directions are East and North. Unit vectors in these directions are denoted by $\mathbf{i}$ and $\mathbf{j}$. A particle of mass 2 kg moves in a horizontal plane. The forces $\mathbf{F}_1 = -2\mathbf{i} - 5\mathbf{j}$ and $\mathbf{F}_2 = 5\mathbf{i} + \mathbf{j}$ act on the particle. Find the acceleration of the particle.

(see pages 171–174)

20 Using Newton's law in vector form to find an unknown force.

In this question, $\mathbf{i}$ and $\mathbf{j}$ are the unit vectors horizontally and vertically upwards, respectively.

A particle of mass 0.5 kg experiences two forces, its weight and the force $\mathbf{F}$. The particle accelerates with acceleration $\mathbf{a} = 1.5\mathbf{i} - \mathbf{j}\,\text{ms}^{-2}$. Find the force $\mathbf{F}$.

(see pages 171–174)

21 Resolving forces

In each case below, write the force in vector form, using the $\mathbf{i}$ and $\mathbf{j}$ vectors shown.

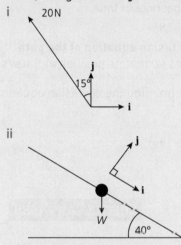

(see pages 176–177)

22 Resolving forces to find acceleration

Annie and Bertie are pushing their car, which has mass 800 kg. Annie pushes with a force of 300 N at an angle of 10° to the direction of travel and Bertie with a force of 200 N at an angle of 15°.

i Show that the resultant force at right angles to the direction of travel is very small.

ii Find the acceleration of the car in the direction of travel.

(see pages 179–180)

23 Analysing motion on an inclined plane

A block of mass 0.1 kg is being pulled by a light string up a smooth plane. The plane makes an angle of 15° with the horizontal and the string is parallel to the slope. The block has an acceleration of 1.5 ms⁻². Find the tension in the string.

(see pages 178–182)

24 Using the model for friction

A block of mass 3 kg rests on a rough horizontal surface. The coefficient of friction between the surface and the block is 0.2. A force of 8 N at 20° to the horizontal pushes the block. Calculate the acceleration of the block.

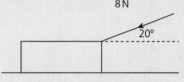

(see pages 202–203)

25 Finding the coefficient of friction

A block of mass 2 kg is placed on a rough plane inclined at 30° to the horizontal. A horizontal force of 20 N acts on the block which is on the point of sliding up the slope. Calculate the coefficient of friction between the block and the plane.

(see pages 202–204)

26 Using moments in an equilibrium problem to find distance

Sally and Nasreen are balancing on a seesaw. Sally has a mass of 25 kg and Nasreen 32 kg. Sally sits 1.4 m from the middle of the seesaw. Calculate the distance from the middle that Nasreen should sit on the other side of the seesaw.

(see pages 188–191)

27 Using moments to solve equilibrium problems

A ruler AB is 30 cm long and has mass 25 g. It is held horizontally by two smooth pegs, one at A and the other at C which is 5 cm from B.

i Assuming that the ruler is uniform, find the contact forces at the pegs.

ii A downwards force is applied at B. Calculate the largest force that can be applied before the ruler tips.

(see pages 188–191)

28 Using moments for problems with a rectangular lamina

A rectangular lamina ABCD is in equilibrium under the action of the forces 3N and 5N at A, 1.5N at B and XN and YN at C as shown in the diagram. The length AB is 30 cm and AD is $d$cm. Calculate the values of X, Y and d.

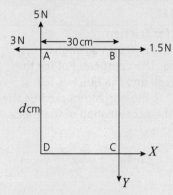

(see pages 188–191)

29 Using differentiation to investigate motion

Abebe runs in a straight line and his displacement in metres from the origin O at time t seconds after the start of his run is given by $s = 15 - 1.47t + 0.01t^3$ for $0 \leqslant t \leqslant 20$.

i   Determine whether his acceleration is constant.

ii Determine whether Abebe changes direction during his 20s run.

(see pages 142–146)

30 Motion with variable acceleration: integration

A particle travels in a straight line and at time ts its acceleration in m s^{-2} is given by $a = 1.5t + 2$. When $t = 0$, the particle is at the origin with a velocity of 4 m s^{-1} in the negative direction.

Find the distance of the particle from the origin when $t = 6$.

(see pages 142–146)

31 Using vectors and calculus to find acceleration

A child runs in a park and his displacement from an origin at time t seconds is given by

$\mathbf{r} = \begin{pmatrix} 0.3t^2 \\ 0.2t^3 + 2t \end{pmatrix}$. Find expressions for the child's

velocity and acceleration. Determine whether the acceleration is constant.

(see pages 184–186)

32 Using vectors to find velocity and displacement

A particle moves with acceleration in m s^{-2} given by $\mathbf{a} = 0.12t\,\mathbf{i} - 0.1\mathbf{j}$. Initially the particle has a velocity of $\mathbf{v}_0 = -\mathbf{i} + 2\mathbf{j}$ and position vector $\mathbf{r}_0 = 5\mathbf{i} - \mathbf{j}$. Find expressions for the velocity and position of the particle at time ts.

(see pages 184–186)

33 Finding the cartesian equation of the path

A particle moves so that its position at time ts

is $\mathbf{r} = \begin{pmatrix} t - 2 \\ 0.1t^3 - 2t \end{pmatrix}$ m. Find the cartesian equation

of the path.

(see pages 184–186)

Short answers on page 224

Full worked solutions online

CHECKED ANSWERS

Chapter 8 Kinematics

About this topic

Kinematics is the study of motion and so is an essential aspect of Mechanics. The language involved often requires the use of technical terms that need to be used precisely, for example the difference between vector and scalar quantities. It is common to supplement descriptions in words with graphs illustrating motion in particular cases, and it is important to be able to draw and interpret such graphs.

In real life, motion is often complicated and it is necessary to use a simplifying model as a first step in analysing it. A very common model involves treating an object as a particle moving along a straight line with constant acceleration. This model can also be used for a large object when its dimensions and any bends in its path are negligible in terms of the overall motion.

You are often seeing things falling, particularly after being thrown up in the air. This common situation can be well modelled as motion in a straight vertical line with a constant acceleration of g downwards due to gravity. Often objects are travelling horizontally as well as vertically and these objects can be modelled as projectiles.

The motion of many objects occurs in 2– or 3–dimensional space and vectors are used to describe and analyse it. There are vector versions of the constant acceleration formulae. It is important to understand the difference between the scalar quantities distance, speed and magnitude of acceleration, which can be calculated from the vector quantities of displacement, velocity and acceleration.

Before you start, remember ...

- The terms **time**, **distance** and **speed** (GCSE level).
- Knowledge of graphical techniques (GCSE level).
- Basic algebra (GCSE level).
- The gradient of a line.
- The formula for calculating the area of trapezium.
- The meanings of **position**, **displacement**, **distance**, **velocity**, **speed** and **acceleration**.
- How to solve equations including simultaneous and quadratic equations.
- Equations of motion in a straight line with constant acceleration (*suvat* equations).

Using graphs to analyse motion

Key facts

1 Vectors and scalars

VECTORS (have magnitude and direction)	SCALARS (have magnitude only)
Displacement	Distance
Position – displacement from a fixed origin	Speed – magnitude of velocity
Velocity – rate of change of position	
Acceleration – rate of change of velocity	
	Time

Warning: although acceleration is strictly a vector, it is commonly used as a scalar. The correct scalar term is magnitude of acceleration.

2 Definitions

- Average speed $= \dfrac{\text{total distance travelled}}{\text{total time taken}}$ (it is a scalar quantity)

- Average velocity $= \dfrac{\text{displacement}}{\text{time taken}}$ (it is a vector quantity)

- Average acceleration $= \dfrac{\text{change in velocity}}{\text{time}}$ (it is a vector quantity)

- Displacement – the distance and direction of one point from another (a vector)
- Position – the displacement from the origin (a vector)
- Distance travelled – the length of the path travelled, whatever the direction (a scalar)

3 Graphs

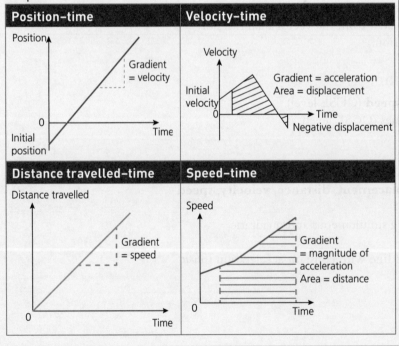

Using graphs

Worked example

Example 1

Gilbert walks North for 40 s at 3.5 m s⁻¹ and then South for 80 s at 2.5 m s⁻¹.
Sketch:
   i   the speed–time graph
   ii the velocity–time graph
   iii the distance travelled–time graph
   iv the position–time graph.

Solution

Choose the origin as Gilbert's starting point and North as the positive direction.

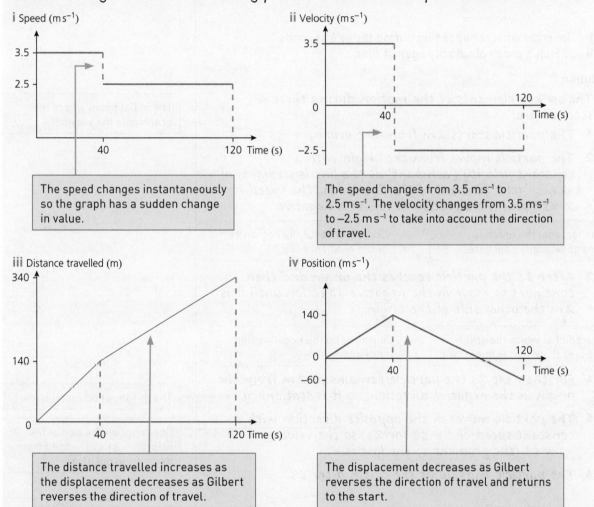

i Speed (m s⁻¹)

The speed changes instantaneously so the graph has a sudden change in value.

ii Velocity (m s⁻¹)

The speed changes from 3.5 m s⁻¹ to 2.5 m s⁻¹. The velocity changes from 3.5 m s⁻¹ to −2.5 m s⁻¹ to take into account the direction of travel.

iii Distance travelled (m)

The distance travelled increases as the displacement decreases as Gilbert reverses the direction of travel.

iv Position (m s⁻¹)

The displacement decreases as Gilbert reverses the direction of travel and returns to the start.

Gilbert covers a total distance of 340 m. Gilbert ends up 60 m south of his starting point.

Hint: The key point here is that distance travelled does not take into account the direction of travel, and so increases even when Gilbert is returning towards the origin.

Worked example

Example 2

The motion of a particle is illustrated by the position–time graph below.

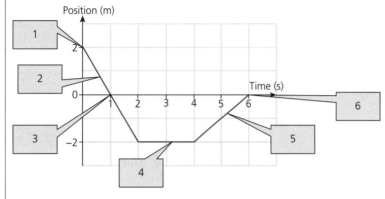

i   Describe what is happening during these 6 seconds.
ii Sketch a graph of velocity against time.

Solution

i   The six key elements of the motion during these 6 seconds are:

> The initial point, where the graph cuts the *y*-axis.

1   The particle starts 2 m from the origin.

2   The particle moves from the origin with a constant velocity ('constant' as the line is straight) of $-2\,\text{ms}^{-1}$ (the gradient of the line is -2). The speed is $2\,\text{ms}^{-1}$ and the direction of travel is negative.

> The fact that the graph is straight is significant here.

> Comment about the negativity of the gradient here.

3   After 1 s the particle reaches the origin and then continues to move in the negative direction until it is 2 m the other side of the origin.

> The point at which the graph crosses the *x*-axis is important.

> Comment on the negative sign for displacement.

4   For the next 2 s the particle remains at 2 m from the origin in the negative direction, so it is stationary.

> The graph is horizontal here.

5   The particle moves in the opposite direction with a constant speed of $1\,\text{ms}^{-1}$ for 2 s, so the velocity is $+1\,\text{ms}^{-1}$ (the gradient of the line is $+1$).

> Comment on the sign of the velocity – it has changed from negative to positive.

6   The particle ends up at the origin after 6 s.

> Make a comment about the final position.

ii

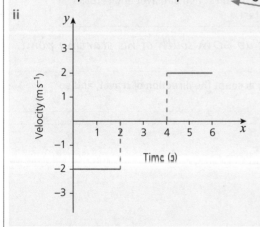

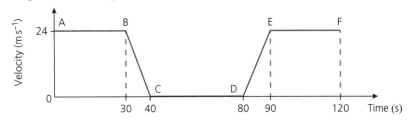

Worked example

Example 3

The velocity–time graph illustrates the progress of a car along a straight road during a 2-minute period.

i   Describe the car's journey.
ii Draw the acceleration–time graph.
iii How far does the car travel?

Solution

i   AB: the car travels at a steady velocity of 24 ms⁻¹ for 30 s.

BC: the car decelerates uniformly to a stop in 10 s.

CD: the car is stationary for 40 s.

DE: the car accelerates uniformly for 10 s reaching a velocity of 24 ms⁻¹.

EF: the car travels at a steady velocity of 24 ms⁻¹ for 30 s.

ii In AB, CD and EF the acceleration is 0 ms⁻².

In BC the acceleration is $\dfrac{-24}{10} = -2.4\,\text{m s}^{-2}$.

In DE the acceleration is $\dfrac{24}{10} = 2.4\,\text{m s}^{-2}$.

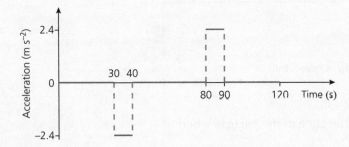

iii The displacement of the car is the same as the actual distance it travels as the velocity is not negative at any stage during the 2 minutes. It is found by calculating the area under the velocity–time graph.

The velocity–time graph is symmetrical, so the area

$= \dfrac{1}{2}(30 + 40) \times 24 \times 2 = 1680$.

The distance travelled is 1680 m.

Worked example

Example 4

A train travels between two stations which are 8.4 km apart. The train accelerates for 40 s before reaching its maximum speed. It then travels at this speed for 5 minutes before being brought to rest in 60 s with a constant deceleration. What is the maximum speed of the train?

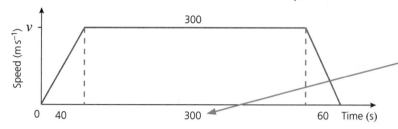

Remember to change minutes into seconds.

Solution

The sketch of the speed–time graph of the journey shows the given information, with suitable units.

The maximum speed is v ms^{-1}.

$$\text{The area is } \frac{1}{2}(400 + 300)v = 8400.$$

Express the distance as 8400 m.

Use the area of the trapezium to represent the total distance travelled.

$$\Rightarrow v = \frac{8400}{350} = 24.$$

The maximum speed of the train is 24 ms^{-1}.

Worked example

Example 5

The velocity of a particle at time t seconds is given by $v = 0.1t^3 - t^2 + 15$ ms^{-1}.

   i   Plot a velocity–time graph for $0 \leqslant t \leqslant 8$.

   ii Use the graph to estimate the acceleration of the particle when $t = 4$.

Solution

i

t	0	1	2	3	4	5	6	7	8
v	15	14.1	11.8	8.7	5.4	2.5	0.6	0.3	2.2

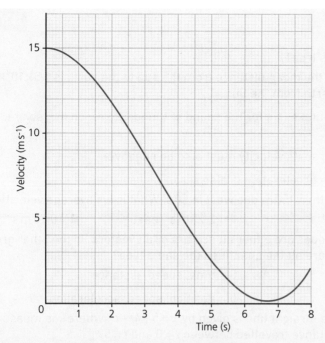

ii Acceleration is the gradient of the velocity–time graph. Draw the tangent to the curve at $t = 4$. ◄——

Choose two points on the tangent that are easy to read accurately and use them to calculate the gradient of the tangent.

> The gradient of the graph is changing. Use the tangent to show the gradient at the required point.

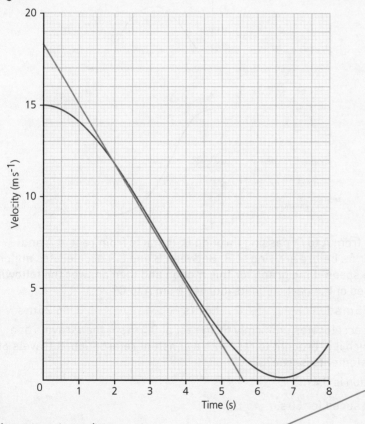

> This method works well when the graph is drawn from a set of values for velocity. In this case, calculus could be used to find the gradient (see Chapter 10).

Points (5, 2) and (1, 15)

$$\text{Gradient} = \frac{15 - 2}{1 - 5} = \frac{13}{-4} = -3.25 \, \text{m s}^{-2}.$$

> **Common mistake:** Take care with the signs here – check by looking at the graph that the gradient is supposed to be negative.

1 Which one of the following statements is true?

A   The speed of light is 3×10^8 m s^{-1} and the mean distance from the sun to the earth is 1.5×10^8 km, so it takes 0.5 s for light to reach the Earth from the Sun.

B   The speed of sound is 340 m s^{-1} and it takes 5 s for the sound of thunder to reach me, so I must be 1.7 km away.

C   A particle has negative acceleration, so its velocity must also be negative.

D   If a particle has zero velocity then its acceleration is also zero.

2 Alan walks 300 m due east in 150 s, and then 150 m due west in 50 s. What is his average velocity?

A   2.25 m s^{-1} B   0.75 m s^{-1} C   2.25 m s^{-1} east   D   0.75 m s^{-1} east

3 The quantities in one of the following groups are either all scalars or all vectors. In the other groups there are some of each. In which group are all the quantities the same type?

A   distance, velocity, acceleration   C   time, speed, distance

B   time, displacement, speed   D   position, speed, acceleration

4 The position of a particle moving along a straight line is given by $x = 5 + 4t - t^2$, where x is measured in metres and t in seconds. What is the distance travelled between $t = 0$ and $t = 5$?

A   −5 m   B   5 m   C   9 m   D   13 m

5 The driver of a car, which has been stopped at traffic lights, finds that she is then stopped at the next traffic lights. Which of the following graphs could represent the velocity–time graph for the car as it travels between the two sets of traffic lights?

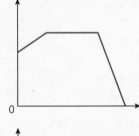

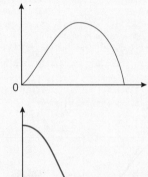

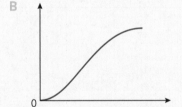

6 A particle moves in a straight line from A to C, passing through B. It starts from rest at A and accelerates uniformly at 1 m s^{-2} for 5 s before arriving at B. Between B and C it accelerates uniformly at 2 m s^{-2} for another 5 s. Draw the speed–time graph for this motion and then answer the following question. What is the average speed of the particle in its journey from A to C?

A $6\frac{2}{3}$ m s^{-1} B   3.75 m s^{-1} C   1.5 m s^{-1} D   6.25 m s^{-1}

7 A body travels in a straight line. It accelerates uniformly from rest at 0.4 m s^{-2}, then travels at a constant speed of 8 m s^{-1}, after which it is brought to rest with a constant retardation. It travels 600 m in 100 s. Which of the following statements is true?

A   The magnitude of the retardation is 0.4 m s^{-2}.

B   The body travels at a constant speed for 50 s.

C   The body travels half the distance in half the time.

D   If the acceleration had been 0.8 m s^{-2} instead of 0.4 m s^{-2}, the body would have travelled for longer at a constant speed.

8 A particle, which starts from rest, travels along a straight line. Its *acceleration* during the time interval $0 \leqslant t \leqslant 10$ is given by this acceleration–time graph. When is the speed of the particle greatest?

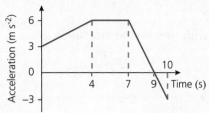

| A | $t=4$ | B | $4 \leqslant t \leqslant 7$ | C | $t=9$ | D | $t=10$ |

9 The graph shows the displacement of a particle at time t. Use the graph to estimate the velocity of the particle at time $t = 4\,\text{s}$.

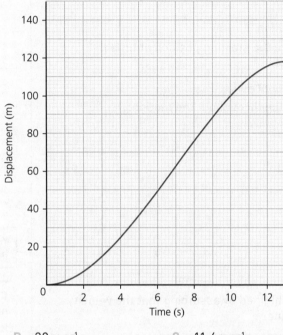

| A | $7.5\,\text{m}\,\text{s}^{-1}$ | B | $30\,\text{m}\,\text{s}^{-1}$ | C | $11.4\,\text{m}\,\text{s}^{-1}$ | D | $9.2\,\text{m}\,\text{s}^{-1}$ |

Full worked solutions online

CHECKED ANSWERS

Exam-style question

P and Q are two points 800 m apart on a straight road. A car passes the point P with a speed of $11\,\text{m}\,\text{s}^{-1}$ and immediately accelerates uniformly, reaching a top speed of $25\,\text{m}\,\text{s}^{-1}$ in 5 s. The driver continues at this speed for another 20 s before decelerating uniformly for T seconds at $1.25\,\text{m}\,\text{s}^{-2}$ until he reaches Q. When the car passes Q its speed is $V\,\text{m}\,\text{s}^{-1}$.

i Sketch the speed–time graph for the journey between P and Q.

ii Find:
  A the acceleration
  B the distance travelled from P to reach the top speed.

iii Find the values of T and V.

iv A Find the average speed for the journey between P and Q.
  B Draw the acceleration–time graph for the journey from P to Q.

Short answers on page 225

Full worked solutions online

CHECKED ANSWERS

Using the constant acceleration formulae

Key facts

1 The *suvat* equations for motion with constant acceleration are:

$$s = \frac{1}{2}(u+v)t$$

$$v = u + at$$

$$s = ut + \frac{1}{2}at^2$$

$$v^2 = u^2 + 2as$$

$$s = vt - \frac{1}{2}at^2$$

a is the constant acceleration.

s is the displacement from the starting position at time t.

v is the velocity at time t.

u is the initial velocity (when $t = 0$).

a and u have the same value throughout the motion.

s and v vary as t varies.

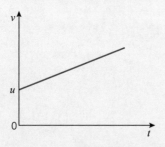

2 The *suvat* equations are obtained by assuming that the velocity–time graph is a straight line.

3 The gradient is the acceleration and this is constant.

> You should be able to derive these formulae.

> **Common mistake:** Remember not to use the *suvat* equations unless you are sure the acceleration is constant.

Worked example

Example 1

A lorry accelerates from rest to $12\,\text{m}\,\text{s}^{-1}$ in $7\,\text{s}$. It then travels at constant speed for $20\,\text{s}$.

 i Calculate the acceleration of the lorry in the first $7\,\text{s}$.

 ii Calculate the total distance travelled.

Solution

i *First decide which equation to use.*

You know $u = 0$, $v = 12$, and $t = 7$. You need a.

Choose $v = u + at$

Now substitute the values you know:

$12 = 0 + a \times 7$

$a = \dfrac{12}{7} = 1.71$

The acceleration is $1.71\,\text{m}\,\text{s}^{-2}$.

> Choose the equation which does not include s.

ii The distance must be found separately for the two phases of motion.

In the first phase, you know $u = 0$, $v = 12$, and $t = 7$. You need s.

Choose $s = \frac{1}{2}(u+v)t$

$s = \frac{1}{2}(0+12) \times 7 = 42$

In the second phase of motion, the velocity is constant.

$s = 12 \times 20 = 240$

Total distance travelled is $42 + 240 = 282\,\text{m}$.

> If you use your value for a, you can use any of the *suvat* equations.

Worked example

Example 2

Starting at $25\,\text{m s}^{-1}$, a car slows down (decelerates) at $2\,\text{m s}^{-2}$.

  i How far has it travelled by the time it comes to rest?

  ii How long does it take to travel $100\,\text{m}$ from its starting point?

Solution

i A deceleration (or retardation) of $2\,\text{m s}^{-2}$ is an acceleration of $-2\,\text{m s}^{-2}$.

You know $u = 25$, $v = 0$, $a = -2$ and you need s.

Choose $v^2 = u^2 + 2as$

$0^2 = 25^2 + 2 \times (-2)s$

$s = \frac{625}{4} = 156.25$

> In this question t is not given nor required so choose the equation that does not involve t.

The distance travelled is $156\,\text{m}$ (correct to 3 significant figures).

ii Now you know $u = 25$, $a = -2$ and $s = 100$, and you need t.

Choose $s = ut + \frac{1}{2}at^2$

Giving $100 = 25t - 1t^2$

$t^2 - 25t + 100 = 0$

> Alternatively, you can use your calculator to solve this equation.

Rearranging $(t - 5)(t - 20) = 0$.

$t = 5$ or $t = 20$

The car had stopped after $12.5\,\text{s}$ so $t = 20$ is rejected. The answer is that the car takes $5\,\text{s}$ to travel $100\,\text{m}$.

Worked example

Example 3

A car driver travels along a straight road. Her motion is modelled using the assumption that her acceleration is constant. She travels from A to B in 4 s and from B to C in the next 4 s. AB is 44 m and BC is 84 m.

 i   Calculate her speed at A and her acceleration.
 ii Calculate the distance travelled according to this model in the next 4 s.
 iii Calculate the speed of the car driver 12 seconds after leaving A.

Solution

 i   There are two unknowns, so set up a pair of simultaneous equations.

   You need to use u for the initial speed at A, so you need to look at the journeys from A to B and from A to C.

   From A to B $s = ut + \dfrac{1}{2}at^2$

   becomes $44 = 4u + \dfrac{1}{2}a \times 4^2$ or $4u + 8a = 44$

   From A to C $s = ut + \dfrac{1}{2}at^2$

   becomes $44 + 84 = 8u + \dfrac{1}{2}a \times 8^2$ or $8u + 32a = 128$

   Solve the simultaneous equations to give $u = 6$ and $a = 2.5$
   If you need a pen and paper method, the equations simplify to give

   $u + 2a = 11$

   $u + 4a = 16$

   Subtract to give $-2a = -5$ so $a = 2.5$
   Substitute into either equation to get $u = 6$

 ii When $t = 12$ s

   $s = ut + \dfrac{1}{2}at^2 = 6 \times 12 + \dfrac{1}{2} \times 2.5 \times 12^2 = 252$

   Distance travelled beyond C is $252 - 128 = 124$ m

 iii At the end of 12 s, the speed of the car driver is

   $v = u + at = 6 + 2.5 \times 12 = 36.$

   36 m s^{-1} is 129.6 km h^{-1} (about 80 mph).
   By now she would be travelling very fast and so it is more likely that she has not maintained the same constant acceleration.

Common mistake: You cannot use the same letter for the initial speed for AB and the initial speed for BC.

Use the solve facility on your calculator if you have it, or multiply the first equation by 2 and subtract from the second equation.

Common mistake: Include a value from your previous work or a new calculation to support your answer. Do not just give a vague answer like 'she would be going too fast'.

Test yourself

Before you start these questions, cover Question 1 and write down the following *suvat* equations from memory. Do all of them before you check your answers.

Make sure you learn any you get wrong and try this test again after doing Question 1.

A	An equation involving s, t, u, v	D	An equation involving a, s, t, v
B	An equation involving a, t, u, v	E	An equation involving a, s, t, u
C	An equation involving a, s, u, v		

1 For each of the following decide which *single* equation is the most appropriate:

A $v = u + at$ B $s = \frac{1}{2}(u+v)t$ C $s = ut + \frac{1}{2}at^2$ D $v^2 = u^2 + 2as$

i   A van travelling at $12\,\mathrm{m\,s^{-1}}$ stops in $4\,\mathrm{s}$. How far does it travel in that time?

ii A bus stops in a distance of $100\,\mathrm{m}$ from a speed of $10\,\mathrm{m\,s^{-1}}$. What is its acceleration?

iii A ball is dropped from a window at a height of $15\,\mathrm{m}$ and accelerates at $9.8\,\mathrm{m\,s^{-2}}$. How long does it take to reach the ground?

iv What is the speed of the ball in **iii** when it has fallen $10\,\mathrm{m}$?

2 A light aircraft lands with a speed of $30\,\mathrm{m\,s^{-1}}$ and takes $20\,\mathrm{s}$ to come to rest. Assuming constant acceleration, three of the following statements are true and one is false. Which one is false?

A The aeroplane travels $300\,\mathrm{m}$ before coming to rest.

B The acceleration of the aeroplane is $-1.5\,\mathrm{m\,s^{-2}}$.

C The aeroplane travels half the distance in the first $10\,\mathrm{s}$.

D The speed of the aeroplane is halved after $10\,\mathrm{s}$.

3 A train accelerates from rest at $0.2\,\mathrm{m\,s^{-2}}$ for $4000\,\mathrm{m}$. How long does it take and how fast is it travelling?

A $200\,\mathrm{s}$, $40\,\mathrm{m\,s^{-1}}$ B $200\,\mathrm{s}$, $20\,\mathrm{m\,s^{-1}}$ C $20\,\mathrm{s}$, $40\,\mathrm{m\,s^{-1}}$ D $20\,\mathrm{s}$, $4\,\mathrm{m\,s^{-1}}$

4 A car starts at rest and accelerates at $2\,\mathrm{m\,s^{-2}}$ for $4\,\mathrm{s}$. It then travels at constant speed for 10 minutes. How far has it travelled in this time?

A $76\,\mathrm{m}$ B $196\,\mathrm{m}$ C $4800\,\mathrm{m}$ D $4816\,\mathrm{m}$

5 Kirsty runs at a constant speed of $8\,\mathrm{m\,s^{-1}}$. She gives a $15\,\mathrm{m}$ head start to Jonathan who starts from rest and has a constant acceleration of $1\,\mathrm{m\,s^{-2}}$. Which equation gives the time at which Kirsty catches Jonathan?

A $8t = t^2$ B $8t = \frac{1}{2}t^2$ C $8t = \frac{1}{2}t^2 + 15$ D $8t + 15 = \frac{1}{2}t^2$

Full worked solutions online

Exam-style question

A motorcycle is travelling along a straight road with an initial speed of $12\,\mathrm{m\,s^{-1}}$ when it starts to slow down at a traffic light. Its deceleration is constant. The motorcycle stops after $3\,\mathrm{s}$, waits for $10\,\mathrm{s}$ and then accelerates uniformly to its original speed of $12\,\mathrm{m\,s^{-1}}$ in $5\,\mathrm{s}$.

i Calculate the distance travelled by the motorcycle during these $18\,\mathrm{s}$.

ii How much less time would the motorcycle have taken to travel this distance if it had maintained the speed of $12\,\mathrm{m\,s^{-1}}$ throughout?

Short answers on page 225

Full worked solutions online

CHECKED ANSWERS ☐

Vertical motion under gravity

Key facts

1 The acceleration due to gravity, $g\,\mathrm{m\,s^{-2}}$, varies around the world. In this book it is assumed that all questions arise in a place where its value is $9.80\,\mathrm{m\,s^{-2}}$ vertically downwards; this is usually written as $9.8\,\mathrm{m\,s^{-2}}$.

2 $v = 0$ at the highest point of the motion.

3 Always draw a diagram and decide in advance where your origin is and which way is positive. $s = 0$ is the origin. Whatever the position, v is positive in the positive direction. a is $+9.8$ when downwards is positive and -9.8 when upwards is positive.

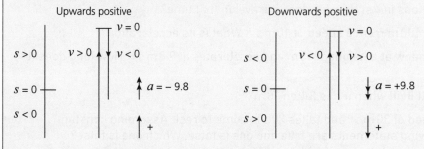

4 In problems where the motion does not begin at the origin so that $s = s_0$ when $t = 0$, replace s in each equation with $(s - s_0)$.

Worked example

Example 1

A ball is projected upwards at $19.6\,\mathrm{m\,s^{-1}}$.
i Find the time it takes to reach its maximum height.
ii Find the actual maximum height.

Solution

i *The ball reaches its maximum height when its velocity is zero.*
First you need to find the time this occurs.
Take upwards as positive and $t = 0$ when it is projected.
You know $u = 19.6$, $a = -9.8$ and $v = 0$ and you need t.

Choose $v = u + at$

$$0 = 19.6 + (-9.8)t$$
$$t = \frac{19.6}{9.8} = 2$$

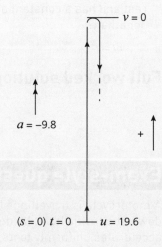

ii *Now take $s = 0$ when $t = 0$.*
To find the maximum height, you need to find s when $t = 2$, $u = 19.6$, $a = -9.8$

$u = 4.9$, $t = 0.5$, $a = 9.8$, $s = ?$

Choose $s = ut + \frac{1}{2}at^2$

$$s = 19.6 \times 2 + \frac{1}{2}(-9.8) \times 2^2 = 19.6\ \mathrm{m}$$

Maximum height is $19.6\,\mathrm{m}$

Worked example

Example 2

A ball is projected vertically upwards from a height of 1 m and reaches its maximum height after 0.5 s. Find:
  i   the initial velocity of the ball
  ii its velocity when it hits the ground.

Solution

i Take upwards to be positive.

You know $v = 0$ when $t = 0.5$ and $a = -9.8$ and you are asked to find u.

Choose $v = u + at$

$$0 = u - 9.8 \times 0.5$$
$$u = 4.9$$

The initial velocity is $4.9\,\text{ms}^{-1}$ upwards.

> **Common mistake:** You are asked for the velocity, so you have to give both the speed $(4.9\,\text{ms}^{-1})$ and the direction (upwards).

ii To find the velocity when the ball hits the ground, you need v when $s = -1$ assuming $s = 0$ when $t = 0$.

You also need to use the value of u that you have just found.

Choose $v^2 = u^2 + 2as$

$$v^2 = 4.9^2 + 2 \times (-9.8) \times (-1)$$
$$v^2 = 43.61$$
$$v = \pm 6.60$$

> The ball is moving downwards so the negative value of v is required.

When the ball hits the ground it has a velocity of $6.60\,\text{ms}^{-1}$ downwards.

> **Common mistake:** It is tempting to say that $v = 0$ when the ball hits the ground because it is likely to stop instantaneously even if it bounces. However, questions like this actually mean 'just before it hits the ground', so $v \neq 0$. The time when $v = 0$ is when it is at maximum height.

Worked example

Example 3

A ball is dropped from a height of 1.6 m, hits the ground and rebounds at half the speed. How high does it bounce?

> It is tempting to assume that the first bounce is half the height, but you might be wrong. It is better to work out the answer properly.

Solution

First find the speed of the ball when it hits the ground.

For this part of the motion,
• take downwards to be positive
• take the point where it is dropped as the origin.

You know $u = 0$, $a = 9.8$, $s = 1.6$, and want to find v.

Choose: $v^2 = u^2 + 2as$
$$v^2 = 0 + 2 \times 9.8 \times 1.6$$
$$v = \sqrt{31.36} = 5.6$$

Rebound speed $= \frac{1}{2} \times 5.6 = 2.8 \text{ ms}^{-1}$

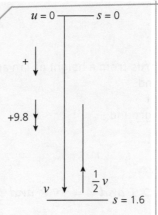

The ball is now travelling upwards, so for this part of the motion,
- take upwards to be positive
- take the point where it bounced as the origin.

$u = 2.8$, $a = -9.8$, $v = 0$ and you need s.

Again, use: $v^2 = u^2 + 2as$
$$0 = 2.8^2 - 2 \times 9.8s$$
$$19.6s = 7.84$$
$$s = 0.4$$

So it bounces to a height of 0.4 m.

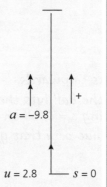

Keep this number in your calculator

The bounce is 0.4 m, so if you thought the bounce would be half of 1.6 m you were wrong.

Worked example

Example 4

A ball is hit upwards and passes the top of a church tower after 2 s and again after 3 s. Find its initial speed, u ms^{-1} and the height of the tower, h m, above the point where the ball is hit.

Solution

Take the ball's starting point to be the origin and upwards to be positive.

The two times when the ball passes the top of the tower are 2 s (on the way up) and 3 s (on the way down).

Now you only know a and two values of t, and you are asked for u and h, so you will need two equations involving u and h.

$a = -9.8$, $u = ?$, $s = h$.

Choose $s = ut + \frac{1}{2}at^2$

On the way up: $h = 2u + \frac{1}{2} \times (-9.8) \times 2^2$
$$h = 2u - 19.6 \qquad (1)$$

On the way down: $h = 3u + \frac{1}{2} \times (-9.8) \times 3^2$
$$h = 3u - 44.1 \qquad (2)$$

Solving equations (1) and (2) simultaneously gives:
$$2u - 19.6 = 3u - 44.1$$
$$u = 24.5$$

Substituting gives $h = 2 \times 24.5 - 19.6 = 29.4$

The initial speed is 24.5 ms^{-1} and the height of the church tower is 29.4 m.

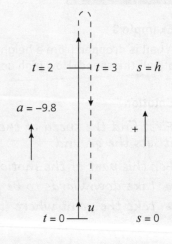

Test yourself

This information applies to Questions 1 to 3.

A ball is thrown upwards from a window with a speed of $3\,\text{m s}^{-1}$ and lands on the ground $15\,\text{m}$ below. Take downwards to be positive and the level outside the window from which the ball is thrown to be the origin.

Three of these statements are false and one is true. Which one is the true statement?

1 A The distance travelled by the ball before it lands is $15\,\text{m}$.

   B The displacement of the ball is positive when it is above the window.

   C The ball's velocity is $+10\,\text{m s}^{-1}$ at some point during the motion.

   D The velocity of the ball is negative when it is below the window.

2 Which one of these is an equation which will give the time, t, taken for the ball to reach the ground.

   A $4.9t^2 - 3t - 15 = 0$ B $4.9t^2 + 3t - 15 = 0$ C $4.9t^2 + 3t + 15 = 0$ D $9.8t^2 - 3t - 15 = 0$.

3 When you have solved the correct equation in Question 2 you can use the results and the symmetry of the motion to give you even more information. Three of the following statements are true and one is false. Solve the correct equation from Question 2 and use your answers to decide which one is false.

   A The ball is at maximum height at $1.04\,\text{s}$.

   B The ball is in the air for $2.08\,\text{s}$.

   C The ball reaches the ground $1.47\,\text{s}$ after passing the window on the way down.

   D The ball passes the window on the way down $0.61\,\text{s}$ after it was thrown up.

   Try to think of another way of working out the correct answer for each statement.

This information applies to Questions 4 and 5.

Starting from rest, a rocket is fired vertically upwards with acceleration $25\,\text{m s}^{-2}$.
After 0.8 seconds there is no more fuel so it continues to move freely under gravity.

4 Calculate the maximum height of the rocket.

   A $8\,\text{m}$ B $16\,\text{m}$ C $20.4\,\text{m}$ D $28.4\,\text{m}$

5 From the moment when the rocket has no more fuel, it takes T s to return to earth. Which of these is a correct equation for T?

   A $4.9T^2 - 20T - 8 = 0$ B $4.9T^2 + 20T - 8 = 0$ C $4.9T^2 - 20T + 8 = 0$ D $4.9T^2 + 20T + 8 = 0$

Full worked solutions online

Exam-style question

A particle, P, is projected vertically upwards at $28\,\text{m s}^{-1}$ from a point O on the ground.
i Calculate the maximum height of P.
When P is at its highest point, a second particle, Q, is projected upwards from O at $25\,\text{m s}^{-1}$.
ii Show that P and Q collide $1.6\,\text{s}$ later and determine the height above the ground that this takes place.

Short answers on page 225

Full worked solutions online

Chapter 9 Variable acceleration

About this topic

You have already studied motion in a straight line where the acceleration is constant. You will now be using calculus to answer similar questions in the more general case where the acceleration need not be constant and may be expressed as a function of time.

Before you start, remember ...

- The language of kinematics and the basic definitions (covered in Chapter 8 of this guide).
- The use of position-time, velocity–time and acceleration–time graphs (in Chapter 8).
- Calculus techniques such as:
  - ○ differentiation to find a gradient and stationary points
  - ○ integration to find the area under a graph.
- Curve sketching techniques.
- The area between a velocity–time curve and the t-axis represents displacement. This is often referred to as 'the area under the graph'. When v is negative, the displacement is negative.
- Areas of the regions between an acceleration-time curve and the t-axis represent changes in velocity; when a is negative the change in velocity is negative.

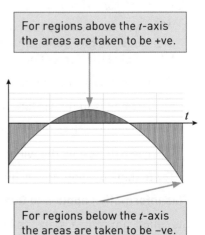

For regions above the t-axis the areas are taken to be +ve.

For regions below the t-axis the areas are taken to be –ve.

Motion using calculus

REVISED

Key facts

1 In this section, the displacement s, the **instantaneous** velocity v and the **instantaneous** acceleration a of a particle moving in a straight line are all taken to be functions of time t.

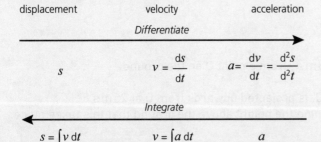

displacement	velocity	acceleration

Differentiate →

$$s \qquad v = \frac{ds}{dt} \qquad a = \frac{dv}{dt} = \frac{d^2s}{d^2t}$$

← *Integrate*

$$s = \int v\, dt \qquad v = \int a\, dt \qquad a$$

When you find velocity from acceleration or displacement from velocity, you use integration and so an arbitrary constant is involved. You need more information to find this constant, but it is not needed if you use definite integration (i.e. integrate between limits).

Common mistake: If you are given an expression for s, v or a in terms of t, only use the constant acceleration (*suvat*) formulae when you are sure that the acceleration is constant.

Hint: Be sure that you know the language of kinematics. Especially important are the distinctions between: *displacement* and *distance travelled*; *velocity* and *speed*. If in the slightest doubt, draw a sketch graph: position–time, velocity–time, distance travelled–time, and so on.

Acceleration

Examples 1–5 are about a particle, P, that moves along the x-axis where the unit of length is the metre. Its velocity, $v\,\mathrm{m\,s^{-1}}$, at time t seconds is given by $v = 0.3t^2 - 2.7t + 4.2$, where $0 \leqslant t \leqslant 10$. When $t = 0$, $x = 3$.

Notice that all the units are S.I.

Worked example

Example 1

Find an expression in terms of t for the acceleration, $a\,\mathrm{m\,s^{-2}}$, of P at time t.

Solution

$v = 0.3t^2 - 2.7t + 4.2$

Differentiating, $a = \dfrac{dv}{dt} = 0.6t - 2.7$

The *suvat* equations cannot be used here as the acceleration is not constant.

Worked example

Example 2

i   Sketch the v–t graph for P for $0 \leqslant t \leqslant 10$.
ii Find the greatest and least values of the velocity of P for $0 \leqslant t \leqslant 10$.

Solution

i   To sketch the quadratic graph, find the minimum point and one or two other points on the graph.

You have seen that $a = \dfrac{dv}{dt} = 0.6t - 2.7$

So $\dfrac{dv}{dt} = 0$ when $t = \dfrac{2.7}{0.6} = 4.5$ and

$v = 0.3 \times 4.5^2 - 2.7 \times 4.5 + 4.2 = -1.875$

When $t = 0$, $v = 4.2$ and when $t = 10$, $v = 7.2$

You could use a table of values instead here, or find the points where the graph crosses axes and plot these.

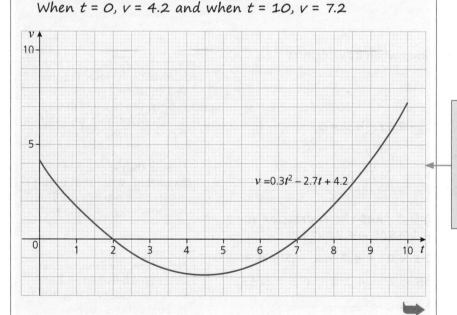

$v = 0.3t^2 - 2.7t + 4.2$

Greatest and least values in an interval are either at the ends or at a maximum or at a minimum point. The graph shows that the greatest value is at the right-hand end-point and that the least value is inside the given interval $0 \leqslant t \leqslant 10$.

ii The greatest value is when $t = 10$ so is given by
$v = 0.3 \times 10^2 - 2.7 \times 10 + 4.2 = 7.2$.
Thus the greatest velocity is $7.2\,\text{ms}^{-1}$ (speed of $7.2\,\text{ms}^{-1}$ in the direction Ox).
For the least value you need $a = 0$.
This occurs when $0.6t - 2.7 = 0$ and so $t = 4.5\,\text{s}$.
When $t = 4.5$, $v = 0.3 \times 4.5^2 - 2.7 \times 4.5 + 4.2 = -1.875$
The least velocity is $-1.875\,\text{ms}^{-1}$ (speed $1.875\,\text{ms}^{-1}$ in the opposite direction to Ox).

> The least value of v occurs when $a = 0$. Since $a = \dfrac{\mathrm{d}v}{\mathrm{d}t}$ this is finding a stationary point.

> The -ve sign shows that the motion is in the opposite direction to Ox.

Position

Worked example

Example 3

Find an expression for the position of P at time t.

Solution

$x = \int (0.3t^2 - 2.7t + 4.2)\,dt$

$\quad = 0.3 \times \dfrac{t^3}{3} - 2.7 \times \dfrac{t^2}{2} + 4.2t + c$

$\quad = 0.1t^3 - 1.35t^2 + 4.2t + c$

Now use the information that $x = 3$ when $t = 0$ to find c.

$\quad 3 = 0.1 \times 0 - 1.35 \times 0 + 4.2 \times 0 + c$

$\quad c = 3$

So $x = 0.1t^3 - 1.35t^2 + 4.2t + 3$.

> x is used because the **position** is being found.

> Remember that $x = 3$ when $t = 0$.

> Put together the expression found above and the value for c to give a final answer.

Worked example

Example 4

What are the positions of P at the times when its velocity is instantaneously zero?

Solution

When $v = 0$, you have $v = 0.3t^2 - 2.7t + 4.2 = 0$.

$\quad 0.3[t^2 - 9t + 14] = 0$

$\Rightarrow 0.3(t - 2)(t - 7) = 0$

$\Rightarrow \qquad\qquad t = 2 \text{ or } t = 7$.

You already know from Example 3 that the expression for the position of P is $x = 0.1t^3 - 1.35t^2 + 4.2t + 3$.

When $t = 2$, $x = 0.1 \times 2^3 - 1.35 \times 2^2 + 4.2 \times 2 + 3 = 6.8$.

When $t = 7$, $x = 0.1 \times 7^3 - 1.35 \times 7^2 + 4.2 \times 7 + 3 = 0.55$.

The positions are $6.8\,\text{m}$ and $0.55\,\text{m}$.

> Use your calculator to solve this equation if you have this facility. Or you could use the quadratic formula, or factorise this to solve the equation.

> Look at the v–t graph in Example 2. When the curve crosses the t-axis, $v = 0$. This happens when $t = 2$ and $t = 7$.

> Notice that the value of x has decreased between $t = 2$ and $t = 7$. The velocity is negative throughout this interval.

Displacement

Displacement is the difference from one position to another. When you use the word, you must also specify the starting position or time; for example, 'the displacement from point A' or 'the displacement from its position at time $t = 0$'. Displacement from the origin is the same as position.

Worked example

Example 5

Find the displacement of P from its position when $t = 0$ to its position when $t = 7$.

Solution

Method 1 Using the position of the particle

You already know from Example 3 that the expression for the position of P is $x = 0.1t^3 - 1.35t^2 + 4.2t + 3$.

When $t = 0$: $x = 3$. ◄———— This is given in the question.

When $t = 7$: $x = 0.1 \times 7^3 - 1.35 \times 7^2 + 4.2 \times 7 + 3 = 0.55$. ◄———— This value was used in Example 4.

So the displacement from its position when $t = 0$ to its position when $t = 7$ is $0.55 - 3 = -2.45\,\text{m}$. ◄

Method 2 Using definite integral

Displacements may be found directly as areas under a graph without finding positions first. Look at the v–t graph for P. This has been shaded to indicate the displacement from $t = 0$ to $t = 7$ with the region above the t-axis shown in green and that below it in orange to emphasise that these represent positive (+) and negative (−) displacements, respectively. The overall displacement is the sum of the **signed** areas of these regions. The process of integration automatically gives the areas the correct signs.

This displacement is negative as the particle finishes at a point nearer to the origin than it started on the positive side of the origin.

$$\int_0^7 v\, dt = \int_0^7 (0.3t^2 - 2.7t + 4.2)\, dt$$

$$= [0.1t^3 - 1.35t^2 + 4.2t]_0^7$$

$$= (0.1 \times 7^3 - 1.35 \times 7^2 + 4.2 \times 7) - (0) = -2.45$$

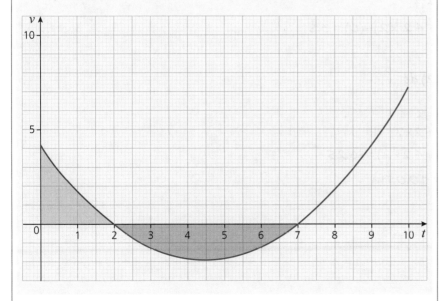

Distance travelled

In Example 6 you will be asked to find the displacement of the particle from $t = 0$ to $t = 7$. In Example 7 you will be asked to find the distance travelled between those times. To understand the difference, look at this diagram. It indicates displacements along the x-axis.

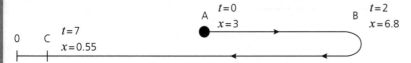

- From $t = 0$ to $t = 2$, the velocity is in the positive direction and the displacement from A to B is positive.
- At $t = 2$, the velocity is zero.
- From $t = 2$ to $t = 7$, the velocity is in the negative direction and the displacement from B to C is negative.

The **displacement** is represented by the journey AC in the negative direction.

The **distance travelled** is the distance AB + the distance BC.

Worked example

Example 6

Find the distance travelled by the particle in the first 7 seconds.

Solution

Method 1 using the positions

The distance AB = $(6.8 - 3) = 3.8$.

The distance BC = $(6.8 - 0.55) = 6.25$.

Total distance = 10.05 m.

Method 2 using the definite integrals

For A to B

$$\int_0^2 v \, dt = \int_0^2 (0.3t^2 - 2.7t + 4.2) \, dt$$
$$= [0.1t^3 - 1.35t^2 + 4.2t]_0^2$$
$$= (0.1 \times 2^3 - 1.35 \times 2^2 + 4.2 \times 2) - (0) = 3.8$$

For B to C

$$\int_2^7 v \, dt = \int_2^7 (0.3t^2 - 2.7t + 4.2) \, dt$$
$$= [0.1t^3 - 1.35t^2 + 4.2t]_2^7$$
$$= (0.1 \times 7^3 - 1.35 \times 7^2 + 4.2 \times 7)$$
$$- (0.1 \times 2^3 - 1.35 \times 2^2 + 4.2 \times 2) = -6.25$$

Total distance = $3.8 + 6.25 = 10.05$ m.

Test yourself

1 A toy is moving in a straight line and its velocity at time t seconds is $v\,\mathrm{m\,s^{-1}}$, where $v = -4t^2 + t + 5$ for $-1 \leqslant t \leqslant 3$. When is the acceleration of the toy zero?

A	$t = 0$	B	$t = 0.125$
C	$t = -0.125$	D	$t = -1$ and $t = 1.25$

2 A particle of grit, G, is stuck to the top of a piece of machinery that is moving up and down a vertical y-axis. The height of G above the ground is y metres at time $t\,\mathrm{s}$ where $y = 10t - 2t^2 - 8$. Determine the direction of motion and speed of G when $t = 3$.

A	Downwards, $2\,\mathrm{m\,s^{-1}}$	B	Downwards, $-2\,\mathrm{m\,s^{-1}}$
C	Upwards, $2\,\mathrm{m\,s^{-1}}$	D	Upwards, $4\,\mathrm{m\,s^{-1}}$

3 A particle is moving in a straight line and t seconds after passing through a point A its velocity is $V\,\mathrm{m\,s^{-1}}$, where $V = 4t - t^2 - 1$ for $0 \leqslant t \leqslant 5$. Draw a velocity–time graph before you start to answer the question. Use it to help you decide which one of the following contains only true statements.

 A Initially the velocity of the particle is $-1\,\mathrm{m\,s^{-1}}$ and its acceleration is $4\,\mathrm{m\,s^{-2}}$ so the displacement of the particle from A after $3\,\mathrm{s}$ is $15\,\mathrm{m}$.

 B The initial velocity of the particle is $-1\,\mathrm{m\,s^{-1}}$ and its velocity after $3\,\mathrm{s}$ is $2\,\mathrm{m\,s^{-1}}$ so the displacement of the particle from A after $3\,\mathrm{s}$ is $1.5\,\mathrm{m}$.

 C The speed of the particle after $1\,\mathrm{s}$ is twice its initial speed; its greatest speed is $3\,\mathrm{m\,s^{-1}}$.

 D The particle does not always travel in the same direction; its greatest velocity is $+3\,\mathrm{m\,s^{-1}}$.

In Questions 4 and 5, an insect is moving along an x-axis. At time t seconds, its velocity is $v\,\mathrm{m\,s^{-1}}$, where $v = 30t - 3t^2 - 63$.

4 Calculate the displacement of the insect from its position when $t = 2$ to its position when $t = 4$.

A	$-2\,\mathrm{m}$	B	$2\,\mathrm{m}$	C	$-150\,\mathrm{m}$	D	$-76\,\mathrm{m}$

5 Calculate the distance travelled by the insect in the time interval $2 \leqslant t \leqslant 4$.

A	$157\,\mathrm{m}$	B	$-2\,\mathrm{m}$	C	$12\,\mathrm{m}$	D	$2\,\mathrm{m}$

Full worked solutions online

Exam-style question

A particle moves along the x-axis with velocity $v\,\mathrm{m\,s^{-1}}$ at time $t\,\mathrm{s}$ given by $v = -5 + 6t - t^2$.

i Find an expression for the acceleration of the particle at time t.

ii Find the times t_1 and t_2, where $t_1 < t_2$, at which the particle has zero velocity.

iii Find the distance travelled between the times t_1 and t_2.

iv At time t_1 the particle passes through the point A. Does the particle pass through A on any later occasion? At time t_2 the particle passes through the point B. Does the particle pass through B on any later occasion?

v Find the distance travelled from $t = 0$ to $t = 6$.

Short answers on page 225

Full worked solutions online

Chapter 10 Forces and Newton's laws of motion

About this topic

Forces are absolutely essential to Mechanics. There are various types of mechanical force and it is essential to know which of them are acting in any situation and to be able to represent them on a diagram, showing their magnitudes and directions, for example when two objects are in contact or connected together.

The basic rules governing the effects of forces are expressed as Newton's laws of motion. Newton's 3rd law, involving action and reaction, is used when two objects are in contact. Newton's 1st and 2nd laws deal with the relationship between force and motion.

Before you start, remember ...

- Forces are vectors. They can be represented by a magnitude (size) and a direction given by an arrow.
- Displacement, velocity, acceleration and force are vectors so each one has a magnitude (size) and a direction.
- Mass, length, speed and time are scalars with magnitude only.
- The *suvat* equations apply to motion with constant acceleration.
- Always draw a force diagram.

Forces

> ### Key facts
>
> 1 Newton's laws of motion:
>
>   1 Every particle continues in a state of rest or of uniform motion in a straight line unless acted on by a resultant external force.
>
>   2 The acceleration of a body is proportional to the resultant force acting on it.
>
>   3 When on object exerts a force on another there is always a reaction of the same kind which is equal, and opposite in direction, to the acting force.
>
> 2 Types of force
>   - Forces on an object due to contact with the surface of another.
>   o **Friction** *always* opposes any tendency to slide.
>   o The **normal reaction** is always perpendicular to the surfaces and to any friction.
>
>   These are the two components of the reaction between the surfaces in contact.

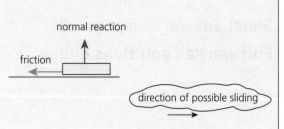

- Forces in a joining rod or string:

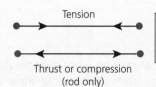

Tension

Thrust or compression
(rod only)

> The tension and compression forces shown act on the objects attached to the ends.

- The tension in a string has the same magnitude on each side of a smooth light pulley but a different direction.

T

T

- The horizontal forces on a wheeled vehicle are usually reduced to three possible forces: resistance, driving force and braking force.
- The **weight** of an object is the force of gravity pulling it towards the centre of the Earth. Weight = mg vertically downwards. The weight of an object is represented by one force acting through its centre of mass.

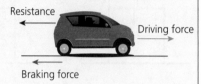

Resistance

Driving force

Braking force

3   Commonly used modelling terms

inextensible	does not vary in length
light	negligible mass
negligible	small enough to ignore
particle	negligible dimensions
smooth	negligible friction
uniform	the same throughout

Worked example

Example 1

Draw diagrams showing the forces acting on:
i   a jug which is being moved along a table by a horizontal force P
ii the table due to the presence of the jug.

Solution

i   *forces on the jug*
ii *forces of the jug on the table*

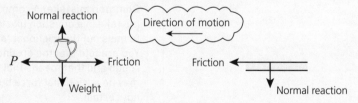

Normal reaction

Direction of motion

P — Friction

Friction

Weight

Normal reaction

> **Common mistakes:** Notice that the weight of the jug and the force P pushing it are present only in the diagram for the jug.

The friction and normal reaction forces acting on the table are equal and opposite to those of the table on the jug.
Because the jug is moving horizontally, the normal reaction between it and the table is equal to its weight. However, the vertical reaction between any two objects is not necessarily equal to the weight of the one on top.

Worked example

Example 2

A pile of three blocks, A, B and C are at rest on a horizontal table. Their weights are W_1, W_2 and W_3, respectively. Draw diagrams to show the forces acting on each of the blocks.

Solution

The diagram shows the forces on the three blocks.

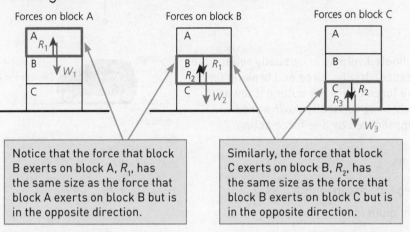

Forces on block A   Forces on block B   Forces on block C

Notice that the force that block B exerts on block A, R_1, has the same size as the force that block A exerts on block B but is in the opposite direction.

Similarly, the force that block C exerts on block B, R_2, has the same size as the force that block B exerts on block C but is in the opposite direction.

Worked example

Example 3

The diagram below shows a sledge on rough horizontal ground. A string, which is in tension, is attached to the front. The sledge is stationary.

Draw a diagram to show the forces acting on the sledge.

String

Sledge

You can simplify the diagram by drawing the sledge as a rectangle.

Solution

The diagram shows the solution.

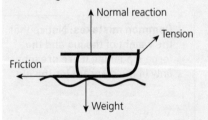

Normal reaction

Tension

Friction

Weight

The tension in the string produces a force to the right (as well as upwards).

You are also told that the sledge is stationary, so there must be a force to the left to oppose it and that has to be friction.

Common mistake: A common mistake in a question like this is to miss out the frictional force. You are told that the ground is rough and that means that there may be a frictional force but it is not certain.

Worked example

Example 4

A particle of mass m kg is on a rough uniform slope that makes an angle of α to the horizontal. It is held at rest by a light string which makes an angle of β with the slope. The tension in the string is T N. The particle is on the point of sliding down the slope.

Draw a diagram showing all the forces acting on the particle.

Notice that you are told that the slope is rough. That tells you that there can be a frictional force.

Solution

There are four forces acting on the particle as shown in the diagram:

● its weight mg

● the tension in the string, T

● the normal reaction of the slope, R

● the frictional force, F, acting to oppose the likely motion.

Worked example

Example 5

A girl attached to a rope is being lowered down a rock face. The rope passes through a smooth pulley at the top of the cliff and is held by a man. The girl has her feet on a ledge and is momentarily at rest.

Draw diagrams to show the forces acting on:

  i the girl
  ii the man.

Solution

i Forces on the girl   ii Forces on the man

All forces are in newtons.

M_2 is his mass.
N_2 is the normal reaction of the ground.
F_2 is the friction between him and the ground.

M_1 is her mass.
T is the tension in the rope.
N_1 is the normal reaction of the ledge.
F_1 is the friction between her and the ledge.

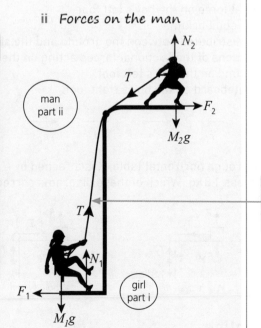

The tension in the rope is the same in both cases because the pulley is smooth.

Test yourself

Questions 1 and 2 are about a box at rest on a rough plane.

1 Which of these is the correct diagram showing the forces acting the box?

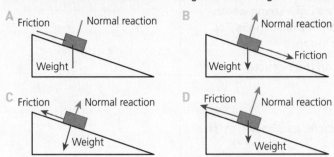

2 Which of these is the correct diagram showing the forces the box exerts on the plane?

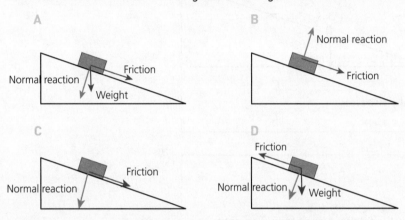

Questions 3 and 4 are about a boy who pushes a skateboard with his right foot while his left foot pushes against level ground. He is gaining speed.

forwards

3 One of the following statements MUST be true, the others may be false. Which one must be true?
  A The vertical reaction between the boy's right foot and the skateboard is equal to his weight.
  B There is a forwards frictional force on the boy's left foot.
  C The forces on the boy are in equilibrium.
  D The boy's weight is equally distributed between the ground and the skateboard.

4 Which diagram shows the directions of the frictional forces acting on the boy?
F_1 is the friction force of the ground on the boy's left foot.
F_2 is the friction force of the skateboard on the boy's right foot.

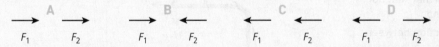

5 A block of mass 5 kg rests on a rough horizontal table. It is attached by a light horizontal string over a smooth pulley to a sphere of mass 1.8 kg. Which of these diagrams correctly shows the forces on the different parts of the system?

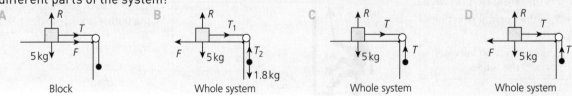

Full worked solutions online

Exam-style question

A courier is delivering a box of books of mass m kg.

First he pushes it along rough level ground with a force P as shown in the diagram.
i Draw a diagram to show the forces acting on the box. Is the normal reaction with the ground equal to the weight of the box?

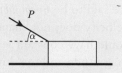

When the courier reaches a ramp, he decides to pull the box with a rope making an angle of β to the horizontal as in the diagram to the right. The ramp also is rough and makes an angle of θ with the horizontal.
ii Draw a diagram to show the forces now acting on the box.

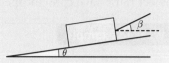

Short answers on page 225

Full worked solutions online

CHECKED ANSWERS

Newton's 1st law of motion

REVISED

Key facts

1   Newton's first law of motion:

Every object continues in a state of rest or uniform motion in a straight line unless it is acted on by a resultant force.

2   Equilibrium:

A particle is in equilibrium when it is stationary or travelling in a straight line at constant speed.
- For forces acting in the same line, the resultant force on the particle is zero.
- For forces given in two perpendicular directions, the resultant force in each direction is zero.
- For forces given in vector notation, the total force is the zero vector.

3   Weight:

The weight of an object is the force of gravity pulling it towards the centre of the Earth. Weight = mg vertically downwards. The weight of an object is represented by one force acting through its centre of mass.

- **S.I. units**

length	metre	m
time	second	s
velocity	metres per second	$m\,s^{-1}$
acceleration	metres per second per second	$m\,s^{-2}$
mass	kilogram	kg
force	newton	N

1 newton is the force required to give a mass of one kilogram an acceleration of $1\,m\,s^{-2}$.

1000 newtons = 1 kilonewton (kN)

Newton's 1st law

When the forces on an object are balanced and have *no resultant* they are said to be *in equilibrium*. Newton's first law says that in that case the velocity does not change. The object could be stationary (at rest) or it could have constant velocity. Constant velocity means that both the magnitude and the direction of the velocity remain unchanged. The object must have constant speed and must be moving in a straight line.

Worked example

Example 1

In each of the following cases state whether the forces on the object are in equilibrium.
i   A motorcycle travels along a straight motorway at 60 m.p.h.
ii An ascending lift approaches the top floor.
iii A cyclist is travelling at steady speed round a circular track.
iv A helicopter is stationary as it hovers over the ground.

Solution

i   The motorcycle has a constant speed and the road is straight so the forces on it must be in equilibrium.
ii The lift must be slowing down as it reaches the top floor so its speed is not constant. The forces are not in equilibrium.
iii The cyclist is not travelling in a straight line so the forces are not in equilibrium.
iv The helicopter is stationary so the forces acting on it are in equilibrium.

> **Common mistake:** Take care **not** to assume that a body is necessarily at rest when the forces acting on it are said to be in equilibrium.

Whenever either its speed or its direction of motion changes, an object has an acceleration. It could be travelling in a straight line with variable speed, or moving with constant speed along a curve, or both speed and direction could be changing.

In order to produce any acceleration, there must be a resultant force acting on the object in the same direction as the acceleration.

Worked example

Example 2

A 2 g leaf is falling at a constant speed. What is the air resistance acting on the leaf?

Solution

The leaf is falling at a constant speed so, assuming this is in a straight line, the forces acting are in equilibrium. The air resistance, R, is equal to the weight of the leaf, mg.

$$\text{Air resistance} = 0.002 \times g \, N$$
$$= 0.0196 \, N$$

> Make sure that the mass is given in kg for the force to be in newtons.

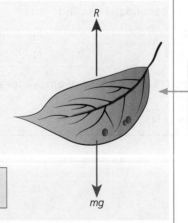

> A diagram showing the forces is useful even in simple cases. You can simplify the drawing to a small circle with two arrows.

Worked example

Example 3

A car of mass 1800 kg is towing a trailer of mass 900 kg at a steady speed along a straight horizontal road.

There is a driving force of D N.

There are also resistance forces of 500 N on the car and 200 N on the trailer.

i Draw clearly labelled diagrams to show the forces on the car and on the trailer.
ii Which of these forces are internal forces when the whole system is considered?
iii Find the value of D.
iv Find the vertical reaction forces and the tension in the towbar.

Solution

i

Forces on trailer   Forces on car

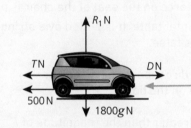

> It is important to attach the forces to the car or the trailer. It matters where the resistances act.

R_1 N and R_2 N are the normal reactions of the road on the car and trailer, respectively. T N is the tension in the towbar.

> Notice that the *weights* of the car and trailer are each given as mg N.

ii The tension, T N, in the towbar is an internal force when the whole system is considered. The force of the car on the trailer is equal and opposite to the force of the trailer on the car (Newton's 3rd law).

iii By Newton's 1st law, the steady speed means that the forces on the whole system are in equilibrium, so

> Consider the car and trailer together as a single object. Make sure you state that the system is in equilibrium.

$$D - 500 - 200 = 0$$
$$D = 700$$

> The resistance forces act in the opposite direction to D and so are negative in this equation. The resultant force is zero.

iv In addition, the forces on each of the car and trailer independently are in equilibrium.

For the car: $R_1 = 1800g$

> Vertical forces balance.

and $D = T + 500$

> Horizontal forces balance.

so $D - 500 - T = 0$
$$T = 200$$

For the trailer: $R_2 = 900g$

> Vertical forces balance.

$$T = 200 \text{ as before.}$$

> Horizontal forces balance.

The reaction forces are 1800g N and 900g N and the tension in the towbar is 200 N.

Test yourself

1 In each of these situations decide whether, during the situation described, the forces acting on the object are in equilibrium:

A always

B never

C some of the time but not all of it

D there is insufficient information to tell.

    i A book is lying on a table.

    ii A sky diver is in free fall before reaching terminal velocity.

    iii A lift in Canary Wharf rises without stopping from ground level to the fiftieth floor.

    iv A child is sitting in a car on a fairground roundabout which is rotating with constant speed.

    v A person is sitting on a bus.

2 Three of the following statements are true and one is false. Which one is false?

A You are in a lift. When it starts moving the force between you and the floor is not equal to your weight.

B A 1.5 kg bag of rice has a weight of 14.7 N.

C When something is moving, there must be a force making it move.

D When you are sitting on a chair, the force on the seat of the chair is normally less than your weight.

3 A small box is in equilibrium on a horizontal table. It is pulled by a string with tension T. Three of the following statements are true and one is false.
Which one is false?

A The friction force, F, is in the wrong direction in the diagram.

B R is the normal reaction of the table on the box.

C $R = mg$

D The magnitude of the tension, T, is greater than the magnitude of F.

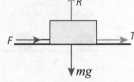

4 A tractor is pulling a trailer with no brakes along level ground using a simple towbar. The tractor brakes suddenly. Draw for yourself a rough sketch showing the forces acting on the tractor and trailer at this moment. Take care to indicate the directions of all the forces.

Three of the following statements are true and one is false. Which one is false?

A The forces are not in equilibrium.

B The tractor and trailer slow down at the same rate.

C The tension in the tow bar must be zero.

D The braking force acts only on the tractor.

Full worked solutions online

CHECKED ANSWERS

Exam-style question

The diagram shows two boxes of mass 8 kg and 6 kg standing together on a rough horizontal surface. Josh tries to push the boxes by applying a horizontal force of 70 N as shown in the diagram. There are frictional forces of $8k$ N and $6k$ N acting on the two boxes, where k is a constant. The boxes do not move.

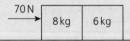

i Draw separate diagrams for each box showing all the forces acting on them.

ii Find the magnitude of the frictional forces acting on the 8 kg box.

iii Find the normal reaction between the 6 kg box and the surface.

iv Find the contact force between the two boxes.

Short answers on page 225

Full worked solutions online

CHECKED ANSWERS

Applying Newton's 2nd law along a line

Key facts

1 Newton's 2nd law:
   - The acceleration of a body is proportional to the resultant force acting on it.
   - For motion in a straight line this is often written as $F = ma$.
   - The acceleration is in the same direction as the resultant force.
   - The equation obtained is often referred to as the 'equation of motion'.
   - Newton's first law is a special case of his second with both F and a equal to zero.
   - Force and acceleration are vector quantities, so the equation can be written **F** = m**a**. See page 172 for more detail.

2 The S.I. units are:

force	newtons
acceleration:	$m\,s^{-2}$

Hint: Acceleration and force are different quantities so it is helpful to use different types of arrow for these: force: $\rightarrow$ acceleration: $\twoheadrightarrow$.

Newton's 2nd Law

A resultant force of $1\,N$ gives a mass of $1\,kg$ an acceleration of $1\,m\,s^{-2}$. This results in the important equation:

$$F = ma$$

Any object falling to Earth has an acceleration of $g\,m\,s^{-2}$ vertically downwards, so the force of gravity acting on it is $mg\,N$. This is its **weight**.

Common mistake: People often talk of things weighing, say, a kilogram when really they mean the mass is 1 kilogram. A 1 kg bag of sugar, for example, has a mass of 1 kg and a weight of $1 \times g = 9.8\,N$.

Worked example

Example 1

A man of mass 75 kg is going up in a lift. Calculate the force between the man and the floor of the lift:

i   when it is accelerating upwards at $0.6\,m\,s^{-2}$
ii when it is moving upwards at a steady speed
iii when it is slowing down at $0.3\,m\,s^{-2}$.

Solution

i *The diagram shows the acceleration and forces acting on the man. Notice that upwards is taken as the positive direction.*

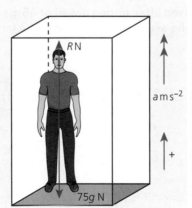

The man's weight is 75gN and the acceleration, a = +0.6.

The resultant force upwards is: R – 75gN.

By Newton's 2nd law:

$$R - 75g = 75 \times 0.6$$
$$R = 45 + 75 \times 9.8$$

The force between the man and the floor is 780N.

ii When the speed is steady, $a = 0$ and the forces on the man are in equilibrium.

Hence the reaction force = the man's weight
$$= 75gN \, (= 735N)$$

iii When the lift is slowing down, $a = -0.3$ so
$$R - 75g = 75 \times (-0.3)$$
$$R = 75 \times 9.8 - 22.5$$
The force between the man and the floor is $712.5N$.

Worked example

Example 2

When a horizontal force of 0.7N is applied to a model car it travels 3.5m in 2s. Calculate the mass of the car assuming it starts from rest and accelerates uniformly across horizontal ground and there is no resistance to its motion.

> The question has a mixture of information about forces as well as distances, etc. Acceleration is the quantity that links them.

Solution

You can calculate the mass of the car using $F = ma$ if you first find the acceleration. You know $u = 0$, $s = 3.5$ and $t = 2$, so use:
$$s = ut + \frac{1}{2}at^2$$

> Choose the equation that does not involve v.

giving $3.5 = 0 + \frac{1}{2}a \times 2^2$
$$3.5 = 2a$$
$$a = 1.75$$
Now use $F = ma$

giving $0.7 = m \times 1.75$
$$m = 0.7 \div 1.75$$
The mass of the car is $0.4\,kg$.

Worked example

Example 3

A racing driver of mass 70kg survived after hitting a wall at $48\,m\,s^{-1}$ and stopping in 0.66m. What was the average magnitude of the force acting on the driver?

> The question gives information about distances and speeds and asks about forces. Find the acceleration to link the two together.

Solution

You are given $u = 48$, $v = 0$ and $s = 0.66$ so the acceleration can be calculated using:
$$v^2 = u^2 + 2as$$
$$0 = 48^2 + 2 \times 0.66 \times a$$
$$-1.32a = 2304$$
$$a = -2304 \div 1.32$$
$$= -1745.45$$

> Choose the equation that does not involve t.

> The car slowed down so the acceleration was negative.

> This is a massive force, but it is based on a true story.

By Newton's second law, the force acting was:
$$mass \times acceleration = 70 \times 1745.45\,N$$
$$= 122181.82\,N \text{ or } 122\,kN$$

> Notice that the rounding is left to the end. Keep the unrounded numbers on your calculator all the way through.

Worked example

Example 4

A tractor of mass 1.7 tonnes exerts a driving force of 3000 N. The resistance to motion is 500 N. Calculate the distance travelled by the tractor by the time it reaches a velocity of 8 m s⁻¹ from rest over horizontal ground.

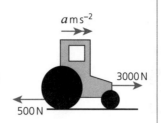

> The question gives information about forces and asks a question about distance. Use acceleration to link these.

Solution

To calculate the acceleration:

Resultant horizontal force = (3000 – 500) N and m = 1700

> Convert the mass of the tractor to kg using 1 tonne = 1000 kg.

So by Newton's second law: $3000 - 500 = 1700a$

The acceleration is $\frac{2500}{1700} = 1.47 \text{ m s}^{-2}$.

You know that $u = 0$, $v = 8$, $a = 1.47$

so choose $\quad v^2 = u^2 + 2as$

> Use the formula that does not contain t.

$8^2 = 0^2 + 2 \times 1.47 \times s$

$s = \frac{64}{2.94} = 21.8$. It travels 21.8 metres.

Test yourself

TESTED

1 Alex and Ben both push a car of mass 850 kg forwards with horizontal forces of 450 N and 420 N, respectively. They produce an acceleration of 0.8 m s⁻². Calculate the resistance to motion.

  A   5794 N   B   650 N   C   190 N   D   710 N

2 In a road safety test a car containing a dummy of mass 60 kg is made to collide with a wall. Initially the car is moving at 16 m s⁻¹ and it buckles when it hits the wall. What average force acts on the dummy if it moves forwards 0.8 m before stopping?

  A   9600 N   B   9012 N   C   1200 N   D   588 N

3 A parachutist of mass 75 kg is descending with a downwards acceleration of 5.2 m s⁻². At this point the upthrust acting on the parachutist is U N. Later, the upthrust has doubled. What is his new acceleration?

  A   2.6 m s⁻²   B   1.67 m s⁻²   C   0.6 m s⁻²   D   0 m s⁻²

4 A block of mass 12 kg is being pushed along a rough horizontal plane by a horizontal force of 24 N. The block starts from rest and when it reaches a speed of 1.5 m s⁻¹ the force is removed. There is a constant frictional resistance to motion of 15 N. The block slows down and comes to rest. For how long from start to finish is the block moving?

  A   1.2 s   B   1.95 s   C   2 s   D   3.2 s

5 You are testing your new bathroom scales in a lift going up several floors without stopping in between. When will the scales show you as heavier than you really are?

  A   Never, they will always be correct.

  B   When the lift starts moving.

  C   Half way up.

  D   When the lift slows down.

Full worked solutions online

CHECKED ANSWERS

Exam-style question

A lorry of mass 40 tonnes is travelling along a straight, level road.

i Calculate the acceleration of the lorry when a resultant force of 60 000 N acts on it in the direction of its motion.

ii How long does it take the lorry to increase its speed from $5\,m\,s^{-1}$ to $12.5\,m\,s^{-1}$?

iii The lorry has an acceleration of $1.3\,m\,s^{-2}$ when there is a driving force of 80 000 N. Calculate the resistance to motion of the lorry.

Short answers on page 225

Full worked solutions online

CHECKED ANSWERS

Connected particles

REVISED

Key facts

1 The acceleration of an object is in the same direction as the resultant force.

2 When Newton's 2nd law is applied to an object, the equation of motion is obtained.

3 The magnitudes of the velocity and acceleration of two objects connected by a rod or a taut inextensible string are always the same.

4 The tension in a string passing over a smooth light pulley is the same on both sides of the pulley.

When two or more objects are connected, there are usually several possible equations of motion you can write down.

When the objects are moving in the same direction, you can consider all of them together as one big object, or you can treat one or more of them separately as separate objects each subject to its own forces.

Always remember that, when the connection is rigid (as is the case in these questions), the speed and acceleration of the objects must always be the same because they are attached.

When a pulley is involved, the directions might be different.

Worked example

Example 1

A locomotive of mass 75 000 kg is pushing a truck of mass 6000 kg into some sidings. The driving force of the locomotive is 2500 N and there are resistances of 250 N on each of the engine and the truck.

i Write down the equations of motion of the whole train, the locomotive and the truck.

ii Calculate the force in the coupling between the locomotive and the truck making it clear whether it is a tension or compression.

Solution

The diagram to the right shows the locomotive pushing the truck towards the right. For the time being, the force in the coupling is shown as a **tension** TN ($\rightarrow\leftarrow$). The acceleration is $a\,ms^{-2}$.

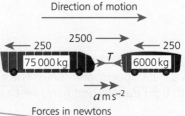

Direction of motion

Forces in newtons

> The diagram shows all the horizontal forces. As the motion is in the horizontal direction, the vertical forces are not needed in equations of motion.

Using $F = ma$ for each separately gives:

i Whole train $2500 - 250 - 250 = 81000a$

   Locomotive: $T + 2500 - 250 = 75\,000a$ (1)

   Truck: $-T - 250 = 6000a$ (2)

> This is an arbitrary choice. Whichever you use, interpret the value of the answer at the end.

> Notice that the resultant force is taken to be in the same direction as the acceleration.

ii Add equations (1) and (2): $2000 = 81000a$

$$a = \frac{2000}{81000} = \frac{2}{81}$$

Substitute for a in equation (2):

$$-T - 250 = 6000 \times \frac{2}{81}$$

$$T = -398.14\ldots$$

> Notice the acceleration can be calculated directly from this equation. This is a good check for your answers from the other two equations.

> If the force in the coupling had been shown as a thrust, it would have turned out to be positive 398 N

The **tension** is negative 398 N indicating that the force in the coupling is in fact a **thrust or compression** ($\leftrightarrow$) of 398 N. It acts forwards on the truck and backwards on the locomotive.

Worked example

Example 2

A boat of mass 1800 kg is pulling a water skier of mass 72 kg. The driving force of the boat's engine is 9000 N. There are resistances of 800 N on the boat and 100 N on the skier.

Calculate the acceleration and the tension in the towrope pulling the skier.

Solution

The diagram to the right shows the horizontal forces acting on the boat and skier.

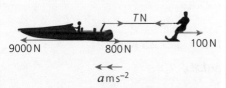

You can calculate the acceleration by considering them both together.

$$(9000 - 800 - 100) = (1800 + 72) \times a$$
$$8100 = 1872a$$
$$a = 8100 \div 1872$$
$$= 4.326\ldots$$

> Resultant force

> Total mass

> Acceleration

To 3 s.f. the acceleration is $4.33\,ms^{-1}$.

To calculate the tension, you need to consider the motion of the boat or the skier alone. For the water skier:

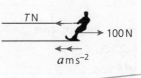

> Notice that the boat and the water skier have the same acceleration.

$$T - 100 = 72a$$
$$T = 72 \times 4.326\ldots + 100$$
$$= 411.538\ldots$$

The tension in the rope is 412 N (to 3 s.f.).

Worked example

Example 3

Blocks of mass 0.5 kg (A) and 0.3 kg (B) are attached by light strings to a block of 0.8 kg (C) which is on a smooth horizontal table. The ends of the light strings pass over smooth light pulleys and hang vertically. The system is released from rest.

i   Draw a diagram to show the forces and acceleration of the blocks.
ii Find the acceleration of the system.
iii Find the tensions in the strings.
iv Find the velocities of the blocks after 2 s.

Solution

i   The forces acting are the weight of each block, the normal reaction of the table on block C and the tensions, T_1 and T_2, in the strings. The block A is heavier so it will accelerate downwards, B will accelerate upwards and C to the left. The acceleration is $a\,\text{ms}^{-2}$. The diagram shows the forces and the accelerations of A, B and C.

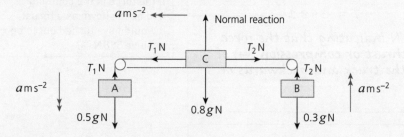

ii Write the equations of motion of A, B and C, each in the direction of its motion.

For A ($\downarrow$):	$0.5g - T_1 = 0.5a$	(1)
For B ($\uparrow$):	$T_2 - 0.3g = 0.3a$	(2)
For C ($\leftarrow$):	$T_1 - T_2 = 0.8a$	(3)

> Notice the vertical forces do not have an effect on the horizontal motion.

When these equations are added, the tensions are eliminated so:

$$0.5g - 0.3g = 0.5a + 0.3a + 0.8a$$
$$0.2g = 1.6a$$

The acceleration is $0.2g \div 1.6 = 1.225\,\text{ms}^{-2}$.

iii Substitute this in (1) to calculate T_1:

$$0.5g - T_1 = 0.5 \times 1.225$$
$$T_1 = 4.2875$$

The tension is 4.2875 N

Substitute this in (2) to calculate T_2:

$$T_2 - 0.3g = 0.3 \times 1.225$$
$$T_2 = 3.3075$$

The tension is 3.3075 N

> Check your answers by substituting your values into equation (3).

iv Take the direction of motion to be positive.

For each block, $u = 0$, $a = 1.225$, $t = 2$.

To calculate v after 2 seconds use $v = u + at$.

$$v = 0 + 1.225 \times 2 = 2.45$$

A has downwards velocity $2.45\,\text{ms}^{-1}$

B is moving upwards at $2.45\,\text{ms}^{-1}$

C is moving to the left at $2.45\,\text{ms}^{-1}$

> The question asks for velocity, not speed, so it is important to include the direction in your answer.

Test yourself

Use the following information for Questions 1 and 2

A car of mass 800 kg is pulling a trailer of mass 600 kg along a straight level road as shown in the diagram above. There is a resistance of 90 N on the car and 200 N on the trailer and they are slowing down at a rate of $0.8\,\mathrm{m\,s^{-2}}$.

1 Calculate the braking force required.
  A 550 N   B 830 N   C 1120 N   D 1410 N

2 By considering the motion of the trailer, calculate the force in the tow bar.
  A Thrust 680 N   B Thrust 280 N   C Tension 200 N   D Tension 280 N

Use the following information for each of the Questions 3 and 4

A block of mass 3 kg is held on a rough table. A light inextensible string attached to the block passes over a smooth pulley and a sphere of mass 2 kg hangs from the other end.

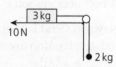

The block is then released and allowed to slide on the table against a friction force of 10 N. The tension in the string is T N and the acceleration of the system is $a\,\mathrm{m\,s^{-2}}$.

3 Draw a diagram for yourself showing the forces acting on the block and the sphere and their accelerations. Use your diagram to write down their equations of motion.
Three of these equations are incorrect and one is correct. Which one is correct?
  A $T = 2a$ B $T - 10 - 3g = 3a$ C $2g - T = 2ga$ D $T - 10 = 3a$

4 Which of the following is the acceleration of the block?
  A $4.8\,\mathrm{m\,s^{-2}}$ B $1.92\,\mathrm{m\,s^{-2}}$ C $9.6\,\mathrm{m\,s^{-2}}$ D $29.6\,\mathrm{m\,s^{-2}}$

5 Two boxes, A of mass 25 kg and B of mass 35 kg, are stacked onto a fork-lift truck with A on top of B. The truck applies an upward force of 800 N to the bottom of box A and both boxes are lifted. Find the magnitude of the contact force between the boxes correct to 3 s.f.
  A 536 N   B 312.5 N   C 361 N   D 313 N

Full worked solutions online

Exam-style question

Blocks of mass 0.5 kg (A) and 0.3 kg (B) are attached to the ends of a light string which hangs vertically over a smooth light pulley. The system is released from rest.
i Draw a diagram to show the forces and acceleration of the blocks.
ii Find the acceleration of the system.
iii Find the tension in the string.
iv Find the velocities of A and B after 2 s.

Short answers on page 225

Full worked solutions online

Chapter 11 Working in 2 dimensions

About this topic

The motion of many objects occurs in 2- or 3-dimensional space and vectors are used to describe and analyse it. It is important to understand the difference between the scalar quantities of distance, speed and magnitude of acceleration, and their vector equivalents of displacement, velocity and acceleration, and the relationships between them.

There are vector versions of the constant acceleration formulae.

Where acceleration is not constant, vectors and calculus can be used together.

Newton's 2nd law of motion can be written in vector form since force and acceleration are vector quantities; the equation also includes mass which is a scalar.

Before you start, remember ...

- The constant acceleration formulae for motion in a straight line (covered in Chapter 8).
- The use of calculus to analyse motion in a straight line with variable acceleration (covered in Chapter 9).
- The use of Newton's 2nd law for motion in a straight line (covered in Chapter 10).
- How to add and subtract vectors.
- How to multiply vectors by a scalar.

Kinematics in 2 dimensions

REVISED

Key facts

1 A vector has magnitude and direction and can be represented by a line segment of an appropriate length and direction indicated by an arrow.

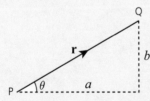

The vector $\mathbf{r} = \overrightarrow{PQ}$ may be written in component form as $a\mathbf{i} + b\mathbf{j}$ or as $\begin{pmatrix} a \\ b \end{pmatrix}$, where $\mathbf{i}$ and $\mathbf{j}$ are unit vectors in the x and y directions.

2 $\mathbf{r} = a\mathbf{i} + b\mathbf{j}$ has a magnitude $\sqrt{a^2 + b^2}$ and direction θ given by $\tan\theta = \dfrac{b}{a}$.

3 A position vector starts at the origin, e.g. $\overrightarrow{OS}$ is a position vector.

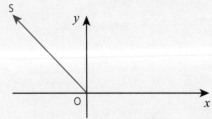

4   Vectors are often written as single letters. When handwritten they should be underlined, e.g. m̲,. When in print, they are in bold type, e.g. **m**.

5   The vector $\overrightarrow{MN}$ in the diagram (from M to N) is given by $\overrightarrow{MN} = \mathbf{n} - \mathbf{m}$.

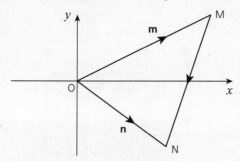

6   **a** and 3**a** are 'like' parallel vectors; **b** and −2**b** are 'unlike' parallel vectors (i.e. in the same direction but in the opposite sense to each other).

7   Vectors in component form are added term by term, i.e. $\begin{pmatrix} a \\ b \end{pmatrix} + \begin{pmatrix} c \\ d \end{pmatrix} = \begin{pmatrix} a + c \\ b + d \end{pmatrix}$.

8   Vectors represented as line segments can be added 'nose to tail' so that the arrows follow round in order. The *resultant* is the vector that goes from the start of the first vector to the end of the last one. If the vectors form a closed polygon then the vector sum is zero.

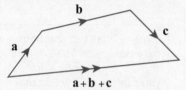

9   A unit vector has a magnitude of 1 in a given direction, e.g. the unit vector in the direction $3\mathbf{i} + 4\mathbf{j}$ is $\frac{3}{5}\mathbf{i} + \frac{4}{5}\mathbf{j}$ (i.e. divide the direction vector by its magnitude).

10 The constant acceleration equations have vector forms:

$$\mathbf{s} = \frac{1}{2}(\mathbf{u} + \mathbf{v})t$$

$$\mathbf{v} = \mathbf{u} + \mathbf{a}t$$

$$\mathbf{s} = \mathbf{u}t + \frac{1}{2}\mathbf{a}t^2$$

$$\mathbf{s} = \mathbf{v}t - \frac{1}{2}\mathbf{a}t^2$$

Worked example

Example 1

Write the vectors shown on the diagram in column vector form.

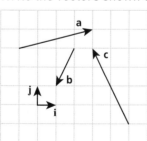

Solution

$a = \begin{pmatrix} 4 \\ 1 \end{pmatrix}, \; b = \begin{pmatrix} -1 \\ -2 \end{pmatrix}, \; c = \begin{pmatrix} -2 \\ 4 \end{pmatrix}$

> Be careful with signs. Take notice of the way the arrow is pointing on the vector.

Worked example

Example 2

Find the magnitude and the direction of the following vectors

i $2\mathbf{i} - 2\mathbf{j}$

ii $-2\mathbf{i} + 2\mathbf{j}$

Solution

i   The magnitude of $2\mathbf{i} - 2\mathbf{j}$ is given by $\sqrt{2^2 + (-2)^2} = 2\sqrt{2}$

and the direction by angle θ so that $\tan\theta = \dfrac{-2}{2} = -1$,

i.e. $\theta = 315°$ (or $-45°$).

> Remember $(-2)^2 = 4$.

> The angle θ is measured from the positive x-axis and in the anticlockwise direction.

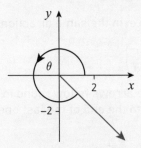

ii The magnitude of $-2\mathbf{i} + 2\mathbf{j}$ is also $2\sqrt{2}$ and the direction θ also has $\tan\theta = -1$ but in this case the solution you want is $\theta = 135°$.

> $\tan\theta = -1$ has two solutions between 0° and 360°; $\theta = 315°$ or $\theta = 135°$ By drawing a diagram you can see clearly that the solution you need in this case is in the fourth quadrant.

> **Hint:** Always draw a diagram to show the examiner exactly which angle you mean.

Worked example

Example 3

Given $\mathbf{a} = \begin{pmatrix} 2 \\ 4 \end{pmatrix}$, $\mathbf{b} = \begin{pmatrix} 3 \\ -2 \end{pmatrix}$ and $\mathbf{c} = \begin{pmatrix} 5 \\ -2 \end{pmatrix}$, find the resultant vectors:

i $\mathbf{a} + \mathbf{b}$

ii $3\mathbf{a} + \mathbf{b} - 2\mathbf{c}$

Solution

i $\mathbf{a} + \mathbf{b} = \begin{pmatrix} 2 \\ 4 \end{pmatrix} + \begin{pmatrix} 3 \\ -2 \end{pmatrix} = \begin{pmatrix} 5 \\ 2 \end{pmatrix}$

ii $3\mathbf{a} + \mathbf{b} - 2\mathbf{c} = 3\begin{pmatrix} 2 \\ 4 \end{pmatrix} + \begin{pmatrix} 3 \\ -2 \end{pmatrix} - 2\begin{pmatrix} 5 \\ -2 \end{pmatrix}$

$= \begin{pmatrix} 6 \\ 12 \end{pmatrix} + \begin{pmatrix} 3 \\ -2 \end{pmatrix} - \begin{pmatrix} 10 \\ -4 \end{pmatrix}$

$= \begin{pmatrix} -1 \\ 14 \end{pmatrix}$

A resultant is obtained when two or more vectors are added together. The algebraic solution to part i is given above. On a diagram this could be represented by adding **a** to **b** 'nose to tail' so that the resultant, **a** + **b**, joins the tail of **a** to the nose of **b**.

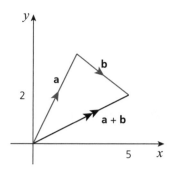

Worked example

Example 4

A, B and C are the points (2, 3), (4, −1) and (−3, −2), respectively.
i   Write down the position vectors of these points in component form.
ii Find the displacements $\overrightarrow{AB}$, $\overrightarrow{BC}$ and $\overrightarrow{CA}$.
iii Show that $\overrightarrow{AB} + \overrightarrow{BC} + \overrightarrow{CA} = 0$ and interpret this result.

Solution

i   Let $\overrightarrow{OA} = \underline{a} = \begin{pmatrix} 2 \\ 3 \end{pmatrix}$, $\overrightarrow{OB} = \underline{b} = \begin{pmatrix} 4 \\ -1 \end{pmatrix}$ and $\overrightarrow{OC} = \underline{c} = \begin{pmatrix} -3 \\ -2 \end{pmatrix}$

ii $\overrightarrow{AB} = \underline{b} - \underline{a} = \begin{pmatrix} 4 \\ -1 \end{pmatrix} - \begin{pmatrix} 2 \\ 3 \end{pmatrix} = \begin{pmatrix} 2 \\ -4 \end{pmatrix}$

$\overrightarrow{BC} = \underline{c} - \underline{b} = \begin{pmatrix} -3 \\ -2 \end{pmatrix} - \begin{pmatrix} 4 \\ -1 \end{pmatrix} = \begin{pmatrix} -7 \\ -1 \end{pmatrix}$

$\overrightarrow{CA} = \underline{a} - \underline{c} = \begin{pmatrix} 2 \\ 3 \end{pmatrix} - \begin{pmatrix} -3 \\ -2 \end{pmatrix} = \begin{pmatrix} 5 \\ 5 \end{pmatrix}$

iii $\overrightarrow{AB} + \overrightarrow{BC} + \overrightarrow{CA} = 2\underline{i} - 4\underline{j} - 7\underline{i} - \underline{j} + 5\underline{i} + 5\underline{j} = 0$

$\overrightarrow{AB}$, $\overrightarrow{BC}$ and $\overrightarrow{CA}$ form a closed polygon when added 'nose to tail'.

> To find $\overrightarrow{AB}$ you need to think about this as a journey from A to O and then from O to B as this takes you along vectors that you know. From A to O is −**a** because this is in the opposite direction to OA. O to B is **b**. So moving from A to B, that is $\overrightarrow{AB}$, is −**a** + **b** = **b** − **a**.

> Notice either notation for vectors can be used.

The next two examples show how vector addition can be used in a physical situation.

Worked example

Example 5

A model aeroplane is set to fly at 12 m s⁻¹ due west. A wind blows at 5 m s⁻¹ due north. Find the resultant velocity of the aeroplane.

Solution

The velocities of the aeroplane and the wind will add together to give the resultant velocity of the aeroplane. First draw a diagram showing this.

> Remember to add nose to tail.

> This angle is the bearing.

By Pythagoras theorem, $R = \sqrt{5^2 + 12^2} = \sqrt{169} = 13$

$$\tan \alpha = \frac{5}{12} \therefore \alpha = 22.619...$$

The resultant velocity of the aeroplane is 13 m s^{-1} on a bearing of 293° (nearest degree).

Worked example

Example 6

A boy wishes to row to a small island that is due west of his current position. He knows that he can row at 10 km h^{-1} in still water and that a current of 3 km h^{-1} flows in a direction of 150°. What course should he steer to reach the island?

Solution

The boy's resultant velocity is in the direction due west. It is the sum of his velocity through the water (10 km h^{-1} in an unknown direction) and the current (3 km h^{-1} at 150°). This information is shown on this diagram.

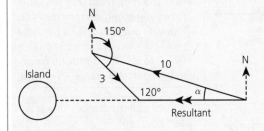

Using the sine rule

$$\frac{\sin 120}{10} = \frac{\sin \alpha}{3}$$

$\therefore \alpha = 15.05°$

The course he needs to steer has a bearing $(270 + 15.05...) = 285°$ (nearest degree).

Examples 7–9 are about a particle, P, with acceleration $(5\mathbf{i} - 3\mathbf{j})$ m s^{-2}. The time is t seconds. When $t = 0$, P is at O with velocity $(-3\mathbf{i} + 4\mathbf{j})$ m s^{-1}. The unit vectors $\mathbf{i}$ and $\mathbf{j}$ have directions east and north, respectively.

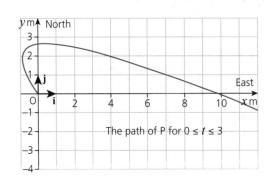

The path of P for $0 \leq t \leq 3$

Worked example

Example 7

Find the speed of P when $t = 3$.

Solution

Since the acceleration is constant (both i and j components are constant), you can use $\underline{v} = \underline{u} + \underline{a}t$ with $t = 3$, $\underline{u} = \begin{pmatrix} -3 \\ 4 \end{pmatrix}$ and $\underline{a} = \begin{pmatrix} 5 \\ -3 \end{pmatrix}$.

> The working is done using column vectors.

This gives $\underline{v} = \begin{pmatrix} -3 \\ 4 \end{pmatrix} + 3 \begin{pmatrix} 5 \\ -3 \end{pmatrix} = \begin{pmatrix} -3+15 \\ 4-9 \end{pmatrix} = \begin{pmatrix} 12 \\ -5 \end{pmatrix}$

> The formula has been used as $u + ta$ as you usually write the scalar multiplier in front of the vector.

The speed of P is the magnitude of v so it is

$\sqrt{12^2 + (-5)^2} = 13\,ms^{-1}$.

Worked example

Example 8

Find the direction of motion of P when $t = 0.6$, giving your answer as a bearing.

Solution

The direction of motion of P is the direction of its velocity.

Using $\underline{v} = \underline{u} + \underline{a}t$ with $t = 0.6$ gives

$\underline{v} = \begin{pmatrix} -3 \\ 4 \end{pmatrix} + 0.6 \begin{pmatrix} 5 \\ -3 \end{pmatrix} = \begin{pmatrix} -3+3 \\ 4-1.8 \end{pmatrix} = \begin{pmatrix} 0 \\ 2.2 \end{pmatrix}$

$\underline{v} = 2.2\underline{j}$ so $\underline{v}$ is in the positive $\underline{j}$ direction which is due north so a bearing of 000°.

Worked example

Example 9

How long after passing through O is P northeast of O?

Solution

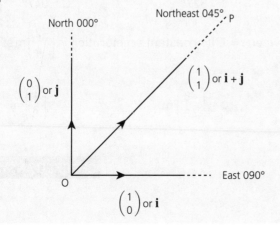

The position of P relative to O is **s**, the displacement from O.

Using $\underline{s} = \underline{u}t + \frac{1}{2}\underline{a}t^2$ gives

$$\underline{s} = t\begin{pmatrix} -3 \\ 4 \end{pmatrix} + \frac{t^2}{2}\begin{pmatrix} 5 \\ -3 \end{pmatrix} = \begin{pmatrix} -3t + 2.5t^2 \\ 4t - 1.5t^2 \end{pmatrix}.$$

When P is northeast of O, the $\underline{i}$ and $\underline{j}$ components of its position are equal and positive.

$-3t + 2.5t^2 = 4t - 1.5t^2$

$\Rightarrow 4t^2 - 7t = 0$

and solving gives $\qquad t(4t-7) = 0$

so $t = 0$ or $t = 1.75$.

P is at O when $t = 0$

and

when $t = 1.75$ P is at $2.40625\underline{i} + 2.40625\underline{j}$ m. The components are equal and positive, so it is northeast of O after 1.75 s.

Test yourself

TESTED

1 A child runs from the front to the back of a bus. The bus is travelling due south at $25\,\text{m s}^{-1}$. The child is running at $0.5\,\text{m s}^{-1}$. What is the resultant velocity of the child?

  A $24.5\,\text{m s}^{-1}$ due S

  B $24.5\,\text{m s}^{-1}$ due N

  C $25.5\,\text{m s}^{-1}$ due N

  D $25.5\,\text{m s}^{-1}$ due S

2 The vector $\mathbf{b} = -8\mathbf{i} + 15\mathbf{j}$. Find its magnitude and direction.

  A 17 units and 61.9° to positive x-axis

  B 17 units and 118.1° to positive x-axis

  C 12.7 units and 118.1° to positive x-axis

  D 17 units and −61.9° to positive x-axis

3 Vectors **a** and **b** are given by $\mathbf{a} = -3\mathbf{i} + \mathbf{j}$ and $\mathbf{b} = \mathbf{i} - 2\mathbf{j}$. Find the magnitude and direction of the resultant of **a** and **b**.

  A 5 units in direction 323.1° to the positive x-axis

  B $\sqrt{5}$ units in direction 26.6° to the positive x-axis

  C $\sqrt{5}$ units in direction 206.6° to the positive x-axis

  D 5 units in direction 143.1° to the positive x-axis

4 A helicopter with a speed in still air of 60 mph needs to fly to a spot Q 80 km away on a bearing of 320° from its current position P. The wind is blowing at 10 mph from the west. What course should the pilot steer?

  A 313°   B 140°   C 327°   D 047°

5 A particle is at $\begin{pmatrix} 0 \\ -1 \end{pmatrix}$ m with velocity $\begin{pmatrix} 2 \\ 3 \end{pmatrix}$ m s^{-1} when $t = 1$. Its constant acceleration is $\begin{pmatrix} -2 \\ 4 \end{pmatrix}$ m s^{-2}.

  Find its position when $t = 4$.

  A $\begin{pmatrix} -3 \\ 26 \end{pmatrix}$ m   B $\begin{pmatrix} 0 \\ 28 \end{pmatrix}$ m   C $\begin{pmatrix} -8 \\ 43 \end{pmatrix}$ m   D $\begin{pmatrix} 6 \\ 8 \end{pmatrix}$ m

Full worked solutions online

CHECKED ANSWERS

Exam-style question

In this question the origin of position vectors is O and **i** and **j** are unit vectors east and north, respectively.

A toy car accelerates uniformly from $\mathbf{u} = -4\mathbf{i}\,\mathrm{m\,s^{-1}}$ to $\mathbf{v} = 2\mathbf{i} + 6\mathbf{j}\,\mathrm{m\,s^{-1}}$ in 3 seconds.
i   Find the acceleration of the car.
ii Determine what bearing the car is travelling on when $t = 1.5\,\mathrm{s}$.
iii Determine the times (if any) when the car is (instantaneously) at rest.

At time $t = 2$, the position vector of the car is $(-\mathbf{i} + 2\mathbf{j})\,\mathrm{m}$.
iv Find the position vector of the car when $t = 0$ and hence an expression for the position of the car at time ts.
v   Determine the times (if any) when the car is southwest of O.

Short answers on page 226

Full worked solutions online

CHECKED ANSWERS

Vector form of Newton's 2nd law

REVISED

Key facts

1 Both force and acceleration are vector quantities. Newton's 2nd law is expressed in vector form as $\mathbf{F} = m\mathbf{a}$.

2 Mass is a scalar quantity.

3 The acceleration is in the same direction as the resultant force.

Worked example

Example 1
i   Find a unit vector in the direction $\begin{pmatrix} 2 \\ -4 \end{pmatrix}$.

ii Hence find a force **F** with magnitude $3\sqrt{5}$ N in the direction $\begin{pmatrix} 2 \\ -4 \end{pmatrix}$.

Solution

$|\mathbf{F}|$ is $3\sqrt{5}$.

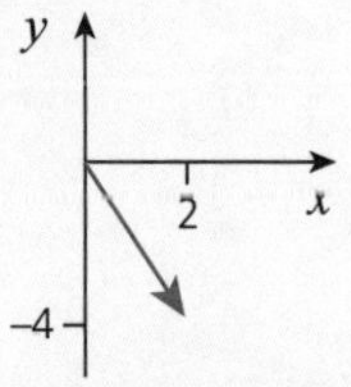

i   Using Pythagoras' theorem (from the diagram above) the magnitude (or length) of $\begin{pmatrix} 2 \\ -4 \end{pmatrix}$ is given by

$$\sqrt{2^2 + (-4)^2} = \sqrt{20} = 2\sqrt{5}$$

So a unit vector in this direction is

$$\frac{1}{2\sqrt{5}}\begin{pmatrix} 2 \\ -4 \end{pmatrix} = \begin{pmatrix} \frac{2}{2\sqrt{5}} \\ \frac{-4}{2\sqrt{5}} \end{pmatrix} = \begin{pmatrix} \frac{1}{\sqrt{5}} \\ \frac{-2}{\sqrt{5}} \end{pmatrix}$$

Hint: Although many questions are set out in $a\mathbf{i} + b\mathbf{j}$ component form, you may find it easier to work in column vector notation, i.e. $\begin{pmatrix} a \\ b \end{pmatrix}$. This can make it easier to spot, or avoid, errors.

ii $\mathbf{F} = 3\sqrt{5} \times \begin{pmatrix} \frac{1}{\sqrt{5}} \\ \frac{-2}{\sqrt{5}} \end{pmatrix} = \begin{pmatrix} 3 \\ -6 \end{pmatrix}$

Notice this is 1.5 multiplied by the given vector.

Worked example

Example 2

In this question, **i** and **j** are unit horizontal and upward vertical vectors, respectively.

A particle of mass 3 kg is acted on by its weight **W** N and forces $\mathbf{F_1}$ N and $\mathbf{F_2}$ N, where $\mathbf{F_1} = 2\mathbf{i} + 4\mathbf{j}$ and $\mathbf{F_2} = -5\mathbf{i} + 2\mathbf{j}$.

i Write the weight of the particle in vector form.

ii Find the acceleration of the particle.

Solution

i *The weight of the particle is 3gN and acts in the negative j direction.*

$$\underline{W} = -3g\underline{j}$$

ii $\underline{W} + \underline{F_1} + \underline{F_2} = -3g\underline{j} + \left(2\underline{i} + 4\underline{j}\right) + \left(-5\underline{i} + 2\underline{j}\right)$

> The resultant force is found by adding the three force vectors.

$$\underline{W} + \underline{F_1} + \underline{F_2} = -3\underline{i} + \left(6 - 3g\right)\underline{j} = m\underline{a}$$

$$\underline{a} = \frac{-3\underline{i} + (6 - 3g)\underline{j}}{3} = \underline{i} - 7.8\underline{j}$$

This solution can also be written in column vector form.

i $\underline{W} = \begin{pmatrix} 0 \\ -3g \end{pmatrix}$

ii $\underline{W} + \underline{F_1} + \underline{F_2} = \begin{pmatrix} 0 \\ -3g \end{pmatrix} + \begin{pmatrix} 2 \\ 4 \end{pmatrix} + \begin{pmatrix} -5 \\ 2 \end{pmatrix} = \begin{pmatrix} -3 \\ 6 - 3g \end{pmatrix}$

$$\underline{a} = \frac{1}{3}\begin{pmatrix} -3 \\ 6 - 3g \end{pmatrix} = \begin{pmatrix} -1 \\ -7.8 \end{pmatrix}$$

Worked example

Example 3

A particle is placed on a smooth horizontal surface. It is in equilibrium under the action of three horizontal forces $\mathbf{F_1} = -2\mathbf{i} + a\mathbf{j}$, $\mathbf{F_2} = a\mathbf{i} + b\mathbf{j}$ and $\mathbf{F_3} = -3a\mathbf{i} + 4\mathbf{j}$.

Find the values of the constants a and b.

Solution

$$\underline{F_1} + \underline{F_2} + \underline{F_3} = 0$$

> The particle is in equilibrium, so the total force is the zero vector.

$$\left(-2\underline{i} + a\underline{j}\right) + \left(a\underline{i} + b\underline{j}\right) + \left(-3a\underline{i} + 4\underline{j}\right) = 0$$

> You can think of this vector equation as two separate equations – one in the **i** direction, and one in the **j** direction.

Look at the i direction.

> Choose the **i** direction first as there is only one unknown in that direction.

$-2 + a - 3a = 0$ *giving* $a = -1$

Look at the j direction.

$a + b + 4 = 0$

> Substitute the value for a.

$-1 + b + 4 = 0$ *giving* $b = -3$

Worked example

Example 4
Two forces $\mathbf{F}_1$ and $\mathbf{F}_2$ act on a particle of mass $1.5\,\text{kg}$. The acceleration of the particle is $\mathbf{a} = -\mathbf{i} + 2\mathbf{j}$.
i If $\mathbf{F}_1 = 3\mathbf{i} - 2\mathbf{j}$, find $\mathbf{F}_2$.
ii If instead $\mathbf{F}_1$ and $\mathbf{F}_2$ are parallel forces but $\mathbf{F}_1$ is twice the magnitude of $\mathbf{F}_2$, find the forces.

Solution
i Newton's 2nd law gives $\underline{F}_1 + \underline{F}_2 = m\underline{a}$

$$\left(3\underline{i} - 2\underline{j}\right) + \underline{F}_2 = 1.5\underline{a} = 1.5\left(-\underline{i} + 2\underline{j}\right)$$

$$\underline{F}_2 = \left(-1.5\underline{i} + 3\underline{j}\right) - \left(3\underline{i} - 2\underline{j}\right)$$

$$\underline{F}_2 = -4.5\underline{i} + 5\underline{j}$$

ii Newton's 2nd law gives $\underline{F}_1 + \underline{F}_2 = m\underline{a}$
$$\underline{F}_1 = 2\underline{F}_2$$

so the equation becomes $2\underline{F}_2 + \underline{F}_2 = 1.5\left(-\underline{i} + 2\underline{j}\right)$

$$3\underline{F}_2 = -1.5\underline{i} + 3\underline{j} \text{ giving } \underline{F}_2 = -0.5\underline{i} + \underline{j}$$

So $\underline{F}_1 = 2\left(-0.5\underline{i} + \underline{j}\right) = -\underline{i} + 2\underline{j}$

> If $\mathbf{F}_1$ and $\mathbf{F}_2$ are parallel forces, they must both be in the same direction as the acceleration.

Worked example

Example 5
A particle of mass $2\,\text{kg}$ accelerates uniformly from $\mathbf{u} = \begin{pmatrix} -3 \\ 1 \end{pmatrix}\,\text{m s}^{-1}$ to $\mathbf{v} = \begin{pmatrix} 7 \\ -4 \end{pmatrix}\,\text{m s}^{-1}$ in $5\,\text{s}$ under the action of two forces $\mathbf{P}$ and $\mathbf{Q}$. The force $\mathbf{P}\,\text{N}$ is given by the vector $\begin{pmatrix} -5 \\ 6 \end{pmatrix}$. Find the magnitude of the force $\mathbf{Q}$.

Solution
To find acceleration when $\underline{u} = \begin{pmatrix} -3 \\ 1 \end{pmatrix}$, $\underline{v} = \begin{pmatrix} 7 \\ -4 \end{pmatrix}$ and $t = 5$.

$$\underline{v} = \underline{u} + \underline{a}t$$

$$\begin{pmatrix} 7 \\ -4 \end{pmatrix} = \begin{pmatrix} -3 \\ 1 \end{pmatrix} + 5\underline{a}$$

$$5\underline{a} = \begin{pmatrix} 7 \\ -4 \end{pmatrix} - \begin{pmatrix} -3 \\ 1 \end{pmatrix} = \begin{pmatrix} 10 \\ -5 \end{pmatrix}$$

$$\underline{a} = \frac{1}{5}\begin{pmatrix} 10 \\ -5 \end{pmatrix} = \begin{pmatrix} 2 \\ -1 \end{pmatrix}$$

> You are told that the acceleration is uniform, so the vector form of the *suvat* equations can be used. Choose the equation that does not involve displacement.

Using Newton's 2nd law
$$\underline{F} = m\underline{a} \text{ becomes } \underline{P} + \underline{Q} = 2\underline{a}$$

> Notice is it necessary to find the vector Q first. Then go on to find its magnitude.

$$\begin{pmatrix} -5 \\ 6 \end{pmatrix} + \underline{Q} = 2 \begin{pmatrix} 2 \\ -1 \end{pmatrix}$$

$$\underline{Q} = 2 \begin{pmatrix} 2 \\ -1 \end{pmatrix} - \begin{pmatrix} -5 \\ 6 \end{pmatrix} = \begin{pmatrix} 9 \\ -8 \end{pmatrix}$$

The magnitude of $\underline{Q}$ is $\sqrt{9^2 + (-8)^2} = \sqrt{145} = 12.0\,N$ (to 3 s.f.)

Worked example

Example 6

In this question, **i** is a horizontal unit vector and **j** is an upward vertical unit vector.

A particle of mass 5 kg is in equilibrium under the action of its weight and two other forces $\mathbf{F_1} = 4a\mathbf{i} + 7\mathbf{j}$ and $\mathbf{F_2} = 8\mathbf{i} + b\mathbf{j}$.
i   Write the weight **W** in the form $\mathbf{W} = x\mathbf{i} + y\mathbf{j}$.
ii Find the values of the constants a and b.

Solution
i   The weight of the particle acts vertically downwards and has no sideways component so it is in the negative $\underline{j}$ direction.

$$\underline{W} = 0\underline{i} - 5g\underline{j} = -49\underline{j}$$

ii The particle is in equilibrium so the total force = 0.

$$\underline{F_1} + \underline{F_2} + \underline{W} = 0$$
$$\left(4a\underline{i} + 7\underline{j}\right) + \left(8\underline{i} + b\underline{j}\right) - 49\underline{j} = 0$$

In the $\underline{i}$ direction $4a + 8 = 0$ so $a = -2$.
In the $\underline{j}$ direction $7 + b - 49 = 0$ so $b = 42$.

Test yourself

TESTED

1 A force of 20 N acts on a particle of mass 6 kg in the direction of the vector $-3\mathbf{i} + 4\mathbf{j}$. Find the acceleration of the particle.

A $-10\mathbf{i} + \dfrac{40}{3}\mathbf{j}\,\mathrm{m\,s^{-2}}$

B $-2\mathbf{i} + \dfrac{8}{3}\mathbf{j}\,\mathrm{m\,s^{-2}}$

C $-0.5\mathbf{i} + \dfrac{2}{3}\mathbf{j}\,\mathrm{m\,s^{-2}}$

D $\dfrac{10}{3}\,\mathrm{m\,s^{-2}}$

2 In this question, **i** and **j** are unit horizontal and upward vertical vectors respectively.
Two forces $\mathbf{F_1} = -\mathbf{i} + 3\mathbf{j}$ and $\mathbf{F_2} = -\mathbf{i} - 2\mathbf{j}$ act on a particle of mass 0.5 kg. Find the acceleration of the particle.

A $-4\mathbf{i} - 7.8\mathbf{j}\,\mathrm{m\,s^{-2}}$

B $-4\mathbf{i} + 2\mathbf{j}\,\mathrm{m\,s^{-2}}$

C $-\mathbf{i} - 1.95\mathbf{j}\,\mathrm{m\,s^{-2}}$

D $8.77\,\mathrm{m\,s^{-2}}$

3 A particle of mass 3 kg moves under the action of a force $\begin{pmatrix} 3.6 \\ -6 \end{pmatrix}$. Initially the particle has a velocity of $\begin{pmatrix} -2.5 \\ 2.6 \end{pmatrix}$. Find the speed of the particle after 5 s.

A $15.3\,\text{m s}^{-1}$

B $\begin{pmatrix} 3.5 \\ -7.4 \end{pmatrix}\,\text{m s}^{-1}$

C $8.19\,\text{m s}^{-1}$

D $31.5\,\text{m s}^{-1}$

4 In this question, **i** and **j** are unit horizontal and upward vertical vectors, respectively.
   A particle of mass 3 kg is acted on by two external forces, $\mathbf{F}_1 = -p\mathbf{i} + 5\mathbf{j}$ and $\mathbf{F}_2 = 2p\mathbf{i} + 4p\mathbf{j}$. The particle moves horizontally. Which of the following statements is not true?

   A The weight of the particle acts in the **j** direction.

   B The **j** component of acceleration is zero.

   C The constant p satisfies the equation $4p + 5 = 3g$.

   D The magnitude of the acceleration is $2.03\,\text{m s}^{-2}$.

5 In this question, **i** and **j** are unit horizontal and upward vertical vectors, respectively.
   The motion of a bird of prey of mass 0.2 kg can be modelled as constant acceleration. Its initial velocity is $1.5\mathbf{i}\,\text{m s}^{-1}$ and 5 seconds later its velocity is $(11.5\mathbf{i} - 54\mathbf{j})\,\text{m s}^{-1}$. The forces on the bird are its weight, the force $\mathbf{F} = 0.8\mathbf{i} - 0.3\mathbf{j}$ N from its wings and the force P N from the wind. Find the force P.

   A $-1.2\mathbf{i} + 0.1\mathbf{j}$ N

   B $-0.4\mathbf{i} - 1.66\mathbf{j}$ N

   C $-0.4\mathbf{i} + 0.1\mathbf{j}$ N

   D $-0.4\mathbf{i} - 3.82\mathbf{j}$ N

6 A particle is in equilibrium under the action of three forces. $\mathbf{F}_1 = 12\mathbf{i} - 5\mathbf{j}$, $\mathbf{F}_2 = 3(\mathbf{i} + \mathbf{j})$ and $\mathbf{F}_3$. Which of the following is the correct expression for $\mathbf{F}_3$?

   A $\mathbf{F}_3 = 15\mathbf{i} - 2\mathbf{j}$

   B $\mathbf{F}_3 = -15\mathbf{i} + 4\mathbf{j}$

   C $\mathbf{F}_3 = -13$

   D $\mathbf{F}_3 = -15\mathbf{i} + 2\mathbf{j}$

Full worked solutions online

CHECKED ANSWERS ☐

Exam-style question

In this question **i** is a unit vector horizontally and **j** is the unit vector vertically upwards.

Forces $\mathbf{F}_1 = a\mathbf{i} + b\mathbf{j}$ and $\mathbf{F}_2 = -5\mathbf{i} - a\mathbf{j}$, act on a particle of mass 0.7 kg.

i   Write the weight of the particle in vector form.
ii Find the values of a and b if the particle is in equilibrium.
iii The values of a and b are 3 and 11, respectively. Find the acceleration of the particle.
iv Initially the particle is at rest. Find the displacement of the particle from its initial position after 5 s.

Short answers on page 226

Full worked solutions online

CHECKED ANSWERS ☐

Resolving forces

Key facts

1 A vector can be resolved into two component vectors. The two vectors are at right angles to each other and this makes it easy to use Pythagoras' theorem and trigonometry.

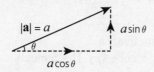
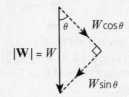

2 If the vector is resolved in **i** and **j** directions then the vector can be written in component form as $\mathbf{a} = a\cos\theta\,\mathbf{i} + a\sin\theta\,\mathbf{j}$.

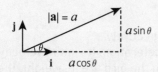

3 When resolving it is important to show the directions in which the vector is to be resolved clearly. Always draw a diagram.

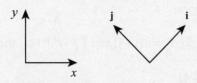

4 If the sum of the **i** components is written as R_1 and that of the **j** components written as R_2 then the resultant vector has a

magnitude of $\sqrt{R_1^2 + R_2^2}$ and is in a direction θ where $\tan\theta = \dfrac{R_2}{R_1}$.

A diagram needs to be drawn to make it very clear which angle is being taken as θ.

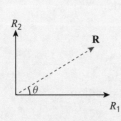

5 In 2 dimensions, Newton's 2nd law is used in its **vector** form $\mathbf{F} = m\mathbf{a}$.

6 The **acceleration** and the **resultant force** are always in the same direction.

7 Problems can often be solved by resolving in two perpendicular directions. This can either be done using vector notation, or by looking at the two directions separately. Make it very clear which direction you are thinking about in each stage of your working.

Worked example

Example 1

Write down the following vectors in component form.

i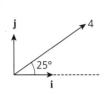

ii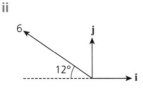

Solution

i $\quad 4\cos25°\underline{i} + 4\sin25°\underline{j}$

$\quad = 3.625...\underline{i} + 1.690...\underline{j}$

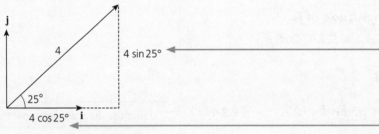

This component is OPPOSITE the angle given.

This component is ADJACENT (next to) the angle given.

ii $\quad -6\sin12°\underline{i} + 6\cos12°\underline{j}$

$\quad = -1.247...\underline{i} + 5.868...\underline{j}$

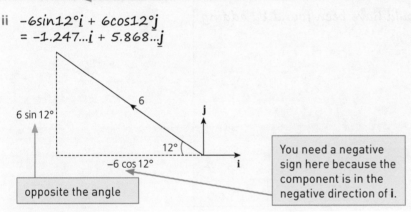

opposite the angle

You need a negative sign here because the component is in the negative direction of **i**.

Hint: Writing a number in the form 3.625... implies that it has not been rounded. Rounding prematurely can lead to inaccurate answers. Always keep as many figures as you can in your calculator while working through a calculation; round off at the end to a suitable number of decimal places or significant figures.

Worked example

Example 2

Forces of magnitude 5 N, 9 N and 3 N act on a body on bearings of 060°, 300° and 180°, respectively.

i Find, in component form, the resultant, **R**, of the three forces.

ii Draw a diagram showing the resultant.

iii Work out the magnitude and direction of the resultant force.

Solution

i *Start by drawing sketches of the three forces. This will help you get the directions right.*

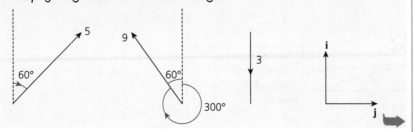

$$5 \sin 60\, \underline{i} + 5 \cos 60\, \underline{j} = 4.330...\underline{i} + 2.5\underline{j}$$
$$-9 \sin 60\underline{i} + 9 \cos 60\underline{j} = -7.794...\underline{i} + 4.5\underline{j}$$
$$-3\underline{j} \qquad\qquad = \qquad\qquad -3\underline{j}$$
$$\overline{\qquad\qquad = -3.464...\underline{i} + 4.0\underline{j}}$$
$$\text{Resultant } \underline{R} = -3.46\underline{i} + 4\underline{j}\ (2\ d.p.)$$

> Resolve each force in turn into $a\mathbf{i} + b\mathbf{j}$ form.

> Add like components together.

ii

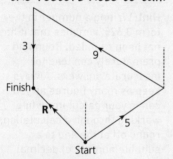

iii Use Pythagoras to find the magnitude of R.
$$|\mathbf{R}| = \sqrt{(-3.46)^2 + 4^2} = \sqrt{27.9716...} = 5.29\,(2\,d.p.)$$

From the diagram $\tan \alpha = \dfrac{4}{3.46}$

Therefore $\alpha = 49.14...°$
The bearing for the resultant is given by $(270° + 49.14...°)$
$= 319°$ (nearest degree).

> Using the diagram from part ii.

Note that the resultant R could have been found by adding the vectors nose to tail.

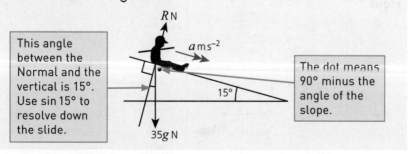

Worked example

Example 3

A boy of mass 35 kg slides down a straight water slide which is 3 m long and inclined at 15° to the horizontal. He starts from rest and the friction is negligible.

i Find his acceleration and his speed when he hits the water at the end of the slide.

ii Find the normal reaction of the slide on the boy.

Solution

i First draw a diagram.

> This angle between the Normal and the vertical is 15°. Use sin 15° to resolve down the slide.

> The dot means 90° minus the angle of the slope.

R N

a ms^{-2}

15°

35g N

There are two forces acting on the boy: his weight $35gN$ vertically down and the normal reaction with the slide RN. There is no friction.

Resolve down the slide. ←————————————

The force RN has no component parallel to the slide. The boy's weight makes an angle of $90° - 15° = 75°$ with the slide, so its component down the slide is $35g\sin15°$ (or $35g\cos75°$)

Using Newton's 2nd law parallel to the slide gives:
$35g\sin15° = 35a$
$$a = 35g\sin15° \div 35$$
$$= 9.8 \times \sin15°$$
$$= 2.536 ...$$

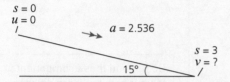

Now you know $u = 0, s = 3, a = 2.536...$ and you need v.
So use $\quad v^2 = u^2 + 2as$
giving $\quad v^2 = 0 + 2 \times 2.536... \times 3 = 15.21...$
$$v = 3.9$$

The boy hits the water at about $3.9\,\text{ms}^{-1}$.

Notice that in this case the mass of the boy (35 kg) cancels out, so the speed should be the same for anyone who uses the slide.

ii To find the normal reaction, resolve perpendicular to the slide. This is perpendicular to the acceleration so the components balance.
$R = 35g\cos15°$ (or $35g\sin75°$)
$\quad = 331.3$
The normal reaction is 331 N.

> The acceleration $a\,\text{ms}^{-2}$ is parallel to the slide, so it is best to resolve parallel and perpendicular to the slide.

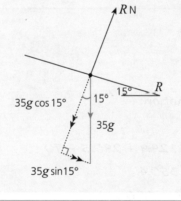

> Make sure that you mark the components of weight differently to the weight itself. Otherwise it looks like the weight is in twice.

> **Hint:** It is usually best to resolve in directions parallel and perpendicular to the direction of the acceleration. Otherwise, you will have to resolve the acceleration as well.

Worked example

Example 4

Two dog teams are pulling a sledge of mass 120 kg with horizontal forces of 150 N and 300 N as shown in the diagram. The sledge is initially at rest. There is a resistance of 70 N opposite to the direction of motion of the sledge.

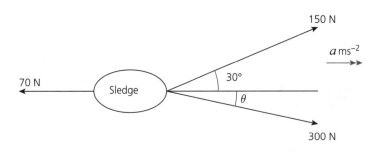

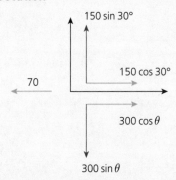

Calculate θ and find the acceleration of the sledge. What is its speed after 3 seconds?

Solution

> Do not add these components to your original force diagram, as that would suggest the components are there in addition to the original force.

The diagram shows the components of the forces in the direction of the acceleration and perpendicular to it.

The resultant force is parallel to the acceleration so the component perpendicular to it is zero.

Resolving perpendicular to the acceleration gives:

$150\sin 30° = 300\sin\theta$

> Use $\sin\theta$ because you are resolving at right angles to the line adjacent to θ.

$150 \times 0.5 = 300\sin\theta$

$0.25 = \sin\theta$

Hence $\theta = 14.48°$

Resolving in the direction of motion:

Resultant force in newtons is:

$150\cos 30° + 300\cos\theta - 70 = 129.9 + 290.5 - 70$

$= 350.4$

Using Newton's 2nd law, $350.4 = 120a$

$$a = \frac{350.4}{120}$$

The acceleration is $2.92\,\text{ms}^{-2}$.

Now you know $a = 2.92$, $u = 0$, and after 3s, $t = 3$.
You need v so use $v = u + at$

$v = 0 + 2.92 \times 3$

$= 8.76$

The speed after 3s is $8.76\,\text{ms}^{-1}$.

Worked example

Example 5

A 3.5 kg block, A, lies on a rough slope inclined at 10° to the horizontal. It is attached to a light inextensible string which passes over a smooth pulley at the top of the slope. Another block, B, of mass 2.5 kg hangs vertically from the end of the string.

The system is held at rest and then released and it is found that the block A takes 2 seconds to slide 1.5 m up the slope.

i Calculate the acceleration of the system.

ii There is a frictional force F N between A and the slope. Find F.

Solution

i

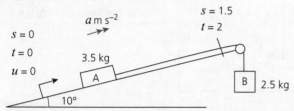

> When solving problems like this you have two strategies for finding an acceleration. One is to use $F = ma$, the other is to use the constant acceleration equations. In this question you are told a lot about the motion, so use the second strategy.

You know $u = 0$, $t = 2$, $s = 1.5$ and you need a.

Use $\qquad s = ut + \dfrac{1}{2}at^2$

$\qquad\qquad 1.5 = 0 + \dfrac{1}{2}a \times 2^2$

giving $\quad 1.5 = 2a$

The acceleration is $0.75\,\text{m s}^{-2}$.

ii

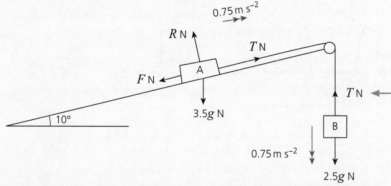

> The blocks accelerate at the same rate because they are joined by an inextensible string.

> First draw a diagram showing all the forces. As well as the weights of the blocks and the friction force F N acting on A, there is a tension force of T N in the string. It is the same throughout because the pulley is smooth.

The acceleration of A is up the slope and the acceleration of B is vertically downwards.

> Because they are moving in different directions, you cannot treat A and B as a single object. Use Newton's 2nd law in the direction of motion for each one separately.

Now you can use $F = ma$.
Resolve vertically for B and parallel to the slope for A.
For B, vertically down: $2.5g - T = 2.5a$ (1)
You know $a = 0.75$ so $2.5g - T = 2.5 \times 0.75$
$$T = 2.5g - 2.5 \times 0.75$$
$$T = 22.625...$$

For A, up the slope:
$$T - F - 3.5g\sin 10° = 3.5a \qquad (2)$$
$$22.625 - 5.956 - 2.625 = F$$
$$14.044 = F$$
The frictional resistance is 14.0 N.

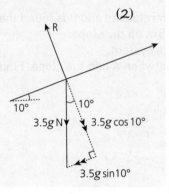

> Alternatively, you can eliminate T from (1) and (2) by adding them and hence find F.

Test yourself

TESTED

1 Find in terms of **i** and **j**, the resultant of the forces shown in the diagram.

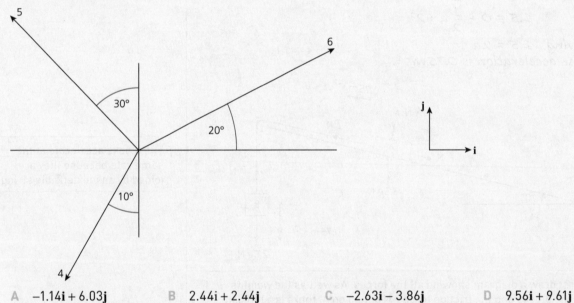

  A $-1.14\mathbf{i} + 6.03\mathbf{j}$ B $2.44\mathbf{i} + 2.44\mathbf{j}$ C $-2.63\mathbf{i} - 3.86\mathbf{j}$ D $9.56\mathbf{i} + 9.61\mathbf{j}$

2 Find the magnitude of the resultant of the vectors shown in the diagram.

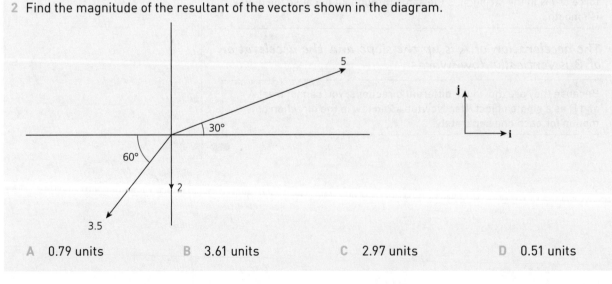

  A   0.79 units   B   3.61 units   C   2.97 units   D   0.51 units

3 The diagram shows a block on a slope.

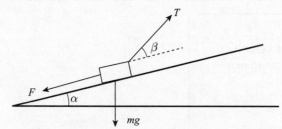

Three of the following statements are false and one is true. Which one is true?

A   The vertical component of T is $T \sin \beta$.

B   The component of F parallel to the slope is $F \cos \alpha$.

C   The component of mg parallel to the slope is $mg \sin \alpha$ down the slope.

D   The component of mg perpendicular to the slope is $mg \sin \alpha$.

Use this information to answer Questions 4 and 5.

A climber of mass 50 kg is being rescued from a ledge using a rope which is pulled through a pulley as shown in the diagram. The pulley is smooth, so the tension in the rope is the same on both sides.

The climber is accelerating upwards at $0.3\,\text{m}\,\text{s}^{-2}$.

Before you answer these questions, draw a diagram showing the tension, TN, in the rope, the weight of the climber and her acceleration.

4 Calculate the tension in the rope

A $T = 65\,\text{N}$ B $T = 490\,\text{N}$ C $T = 505\,\text{N}$ D $T = 637\,\text{N}$

5 Find the magnitude and direction of the force of the rope on the pulley. Three of these statements are true and one is false. Which one is false?

A   Its horizontal component is 414 N.

B   Its vertical component is 795 N.

C   The resultant force is 466 N.

D   The resultant makes an angle of 62.5° with the downward vertical.

Full worked solutions online

CHECKED ANSWERS

Exam style question

A block of mass 12 kg slides down a rough slope which is inclined at 5° to the horizontal.

The block starts with a speed of $15\,\text{ms}^{-1}$ at the top of the slope and it is assumed that there is a constant resistance to motion of 15 N.

i   Calculate the acceleration of the block.

ii For how long does the block slide down the slope? How far does it travel?

Measurements show that the block actually comes to rest in 3.5 s. There is an error caused by assuming the resistance is 15N.

iii Calculate the true value of the resistance.

Short answers on page 226

Full worked solutions online

CHECKED ANSWERS

Calculus and vectors

Key facts

1 In this section, the displacement **s**, the **instantaneous** velocity **v** and the **instantaneous** acceleration **a** of a particle moving in a straight line are all taken to be functions of time t.

displacement velocity acceleration

Differentiate →

$$\mathbf{s} = x\mathbf{i} + y\mathbf{j} = \begin{pmatrix} x \\ y \end{pmatrix} \qquad \mathbf{v} = \frac{d\mathbf{s}}{dt} = \dot{x}\mathbf{i} + \dot{y}\mathbf{j} = \begin{pmatrix} \dot{x} \\ \dot{y} \end{pmatrix} \qquad \mathbf{a} = \frac{d\mathbf{v}}{dt} = \ddot{x}\mathbf{i} + \ddot{y}\mathbf{j} = \begin{pmatrix} \ddot{x} \\ \ddot{y} \end{pmatrix}$$

← *Integrate*

$$\mathbf{s} = \int \mathbf{v}\, dt \qquad\qquad \mathbf{v} = \int \mathbf{a}\, dt \qquad\qquad \mathbf{a}$$

> The dot notation means the time derivative.

2 The direction of the motion of a particle is the direction of its velocity and **not** its position.

3 The cartesian equation of the path is found by equating the components of the position vector to x and y and then eliminating t from the equations.

Worked example

Example 1

The position vector of a bee, **r** m, at time t s is given by $\mathbf{r} = (6t^2 + 3)\mathbf{i} + (2t - 3)\mathbf{j}$. Is the velocity ever zero?

> For a vector to be zero, **both** its components must be **simultaneously** zero.

Solution

$$\mathbf{r} = \begin{pmatrix} 6t^2 + 3 \\ 2t - 3 \end{pmatrix}$$

Now $\mathbf{v} = \dot{\mathbf{r}} = \begin{pmatrix} 12t \\ 2 \end{pmatrix}$, hence **v** is never zero.

> As $x = 6t^2 + 3 \Rightarrow \dot{x} = 12t$ and $y = 2t - 3 \Rightarrow \dot{y} = 2$.
> (Note how the 'dot' notation saves space.)

> Look at **v**. The **i** component is zero when $t = 0$, the **j** component is never 0.

Worked example

Example 2

A small bird has velocity $\begin{pmatrix} 2 \\ 7 \end{pmatrix}$ m s^{-1} when $t = 1$. Its acceleration, **a** m s^{-2}, is given by $\mathbf{a} = \begin{pmatrix} 2t + 4 \\ 3t^2 - 2t + 3 \end{pmatrix}$ for $0 \leqslant t \leqslant 2$, where t is the time in seconds.

Find an expression for the velocity, **v** m s^{-1} of the bird at time t. Find also the velocity of the bird when $t = 2$.

Solution

Since $\mathbf{v} = \int \mathbf{a}\, dt$, $\mathbf{v} = \begin{pmatrix} \int (2t + 4)\, dt \\ \int (3t^2 - 2t + 3)\, dt \end{pmatrix}$

> Notice that integrating **a** means that you integrate each component separately. Each component has an arbitrary constant and you must not assume they are the same.

Hence $\underline{v} = \begin{pmatrix} t^2 + 4t + C \\ t^3 - t^2 + 3t + D \end{pmatrix}$

Using the information that $\underline{v} = \begin{pmatrix} 2 \\ 7 \end{pmatrix}$ when $t = 1$ gives

$\begin{pmatrix} 2 \\ 7 \end{pmatrix} = \begin{pmatrix} 1^2 + 4 \times 1 + C \\ 1^3 - 1^2 + 3 \times 1 + D \end{pmatrix} = \begin{pmatrix} 5 + C \\ 3 + D \end{pmatrix}$ so

$\begin{cases} 2 = 5 + C \text{ and } C = -3 \\ 7 = 2 + D \text{ and } D = 4 \end{cases}$

Hence $\underline{v} = \begin{pmatrix} t^2 + 4t - 3 \\ t^3 - t^2 + 3t + 4 \end{pmatrix}$.

When $t = 2$, $\underline{v} = \begin{pmatrix} 2^2 + 4 \times 2 - 3 \\ 2^3 - 2^2 + 3 \times 2 + 4 \end{pmatrix} = \begin{pmatrix} 9 \\ 14 \end{pmatrix}$ so the velocity is $\begin{pmatrix} 9 \\ 14 \end{pmatrix}$ ms^{-1}.

The path of an object

Worked example

Example 3

The position vector of a particle at time t is given by $\mathbf{r} = (t-1)\mathbf{i} + (t - 2t^2 + 3)\mathbf{j}$. Find the cartesian equation of its path and plot the graph for $-1 \leqslant x \leqslant 3$.

Solution

Referred to the axes Ox and Oy, $\underline{r} = \boldsymbol{x}\underline{i} + y\underline{j}$.

In this question, $\underline{r} = (t - 1)\underline{i} + (t - 2t^2 + 3)\underline{j}$ and so $x = t - 1$ and $y = t - 2t^2 + 3$.

You now eliminate t.

Make t the subject of $x = t - 1$, giving $t = x + 1$.

> The expression for x is simpler than that for y, and so is the one used to make t the subject.

Substituting for t in $y = t - 2t^2 + 3$ gives

$y = (x + 1) - 2 \times (x + 1)^2 + 3$

> This is the cartesian equation of the curve.

$= x + 1 - 2(x^2 + 2x + 1) + 3 = 2 - 3x - 2x^2$

so $y = 2 - 3x - 2x^2$.

Substituting values for x gives the points $(-1, 3)$, $(0, 2)$, $(1, -3)$ and $(2, -12)$. Now plot the graph.

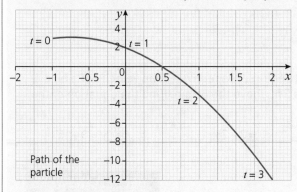

Path of the particle

An important and easy alternative way to obtain this graph is to plot coordinates (x, y) for different values of t.

t	$x = t - 1$	$y = t - 2t^2 + 3$	(x, y)
0	−1	3	(−1, 3)
1	0	2	(0, 2)
2	1	−3	(1, −3)
3	2	−12	(2, −12)

Hint: You can use this method when you cannot make t the subject of either of the expressions for x and y.

Test yourself

1 A boat has position vector, **r** km, at time t hours given by

$\mathbf{r} = \begin{pmatrix} 2t^3 - 9t^2 + 12t + 3 \\ 3t^2 - 2t^3 - 3 \end{pmatrix}$, where $\begin{pmatrix} 1 \\ 0 \end{pmatrix}$ and $\begin{pmatrix} 0 \\ 1 \end{pmatrix}$ are in the directions east and north, respectively, for

$0 \leqslant t \leqslant 2$. The origin of position vectors is at O, the position of a small buoy.
Which of the following statements about the motion of the boat is (completely) true?

A   The time $t = 1$ is not the only time when it is stationary.

B   When $t = 2$ it is travelling southeast.

C   When $t = 0$ it is $3\sqrt{2}$ km from O, travelling east.

D   When $t = 2$ it is northwest of O.

2 The velocity of a particle at time t (in m s^{-1}) is given by $\begin{pmatrix} 0.3t^2 - 0.5t \\ 0.8t \end{pmatrix}$. Initially the particle is at a point

with position vector $\begin{pmatrix} -1 \\ 5 \end{pmatrix}$ m from a fixed origin O. Find the distance of the particle from the origin when $t = 4$ s.

A $\begin{pmatrix} 1.4 \\ 11.4 \end{pmatrix}$ m   B   6.84 m   C   11.5 m   D $\begin{pmatrix} 2.4 \\ 6.4 \end{pmatrix}$ m

3 The acceleration of a toy car, **a** cm s^{-2}, is given by $\mathbf{a} = \begin{pmatrix} 2 \\ 6t^2 - 2t + 1 \end{pmatrix}$ at time t seconds. Its velocity is

$\begin{pmatrix} 3 \\ -1 \end{pmatrix}$ cm s^{-1} when $t = 2$. Find its velocity when $t = 3$.

A $\begin{pmatrix} 6 \\ 48 \end{pmatrix}$ cm s^{-1} B $\begin{pmatrix} 5 \\ 48 \end{pmatrix}$ cm s^{-1} C $\begin{pmatrix} 5 \\ 37 \end{pmatrix}$ cm s^{-1} D $\begin{pmatrix} 5 \\ 20 \end{pmatrix}$ cm s^{-1}

4 A particle is at $\begin{pmatrix} 0 \\ -1 \end{pmatrix}$ m with velocity $\begin{pmatrix} 2 \\ 3 \end{pmatrix}$ m s^{-1} when $t = 1$. Its constant acceleration is $\begin{pmatrix} -2 \\ 4 \end{pmatrix}$ m s^{-2}. Find its
position when $t = 4$.

A $\begin{pmatrix} -3 \\ 26 \end{pmatrix}$ m   B $\begin{pmatrix} 0 \\ 28 \end{pmatrix}$ m   C $\begin{pmatrix} -8 \\ 43 \end{pmatrix}$ m   D $\begin{pmatrix} 6 \\ 8 \end{pmatrix}$ m

5 In this question **i** and **j** are horizontal and vertical unit vectors, respectively. The velocity of a bird at time t s is given by $0.2\mathbf{i} + 0.6t^2\mathbf{j}$ m s^{-1}. Initially the bird is 3 m above the origin. Find the equation of the path of the bird.

A $y = 25x^3$ B $y = 0.2(5x - 15)^3$ C $y = 3 + 25x^3$ D $y = 3 + 15x^2$

Full worked solutions online

Exam-style question

In this question $\mathbf{i}$ and $\mathbf{j}$ are unit vectors East and North.

A model boat moves in a horizontal plane.

At time t, the velocity of the boat is given by $\mathbf{v} = bt^2\mathbf{i} + 4t\mathbf{j}$

where b is a positive constant.

At $t = 1$ second, the magnitude of the acceleration of the boat is $5\,\text{m s}^{-2}$.

i   Show that $b = 1.5$.

At $t = 0$, the boat is at the origin O.

ii Work out the distance of the boat from the origin after the first three seconds of its motion and the bearing of the boat from the origin at that time.

Short answers on page 226

Full worked solutions online

CHECKED ANSWERS ☐

Chapter 12 Moments of forces

About this topic

The modelling assumption that objects are particles with no size is not suitable in all situations. Some objects such as a plank resting on two supports cannot be thought of as a particle and you need to understand the turning effect or moment of the forces. The size and shape of the object and points of application of the forces become an important part of the model.

Before you start, remember ...

- Forces are vectors. They can be represented by a magnitude (size) and a direction given by an arrow.
- Drawing force diagrams – remember to put the arrows through the correct point at which the force acts.
- The weight of an object acts through its centre of mass.
- The conditions for a particle to be in equilibrium.
- How to resolve forces.

Moments

> ### Key facts
>
> 1 The moment of a force F about a point O is given by the product Fd where d is the perpendicular distance from O to the line of action of the force.
>
>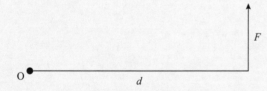
>
> 2 The S.I. unit for the moment of a force is the newton metre (N m).
>
> 3 The usual convention is to take anticlockwise as a positive moment.
>
> 4 A rigid body is only in equilibrium if:
>   - there is no resultant force acting on it
>   - there is no resulting moment acting on it.
>
> 5 When solving a problem, you can take moments about any point, but try to find the point that gives the simplest calculations.

The particle model and the rigid body model

In the particle model, an object is represented by a single particle. You are not concerned with turning forces because the object has no dimensions.

In the rigid body model, an object is represented by a shape, with dimension(s). You need to consider the turning forces on the object.

Worked example

Example 1

In the diagram, the point G is 15 cm from AB and 15cm from CD.

Forces of 4 N act along AB and CD

Find the total moment about G.

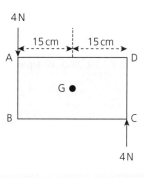

Solution

Moment of the force along AB about G
= 4 × 0.15 = 0.6 Nm anticlockwise

Taking anticlockwise to be the positive direction.

15 cm = 0.15 m

Moment of the force along CD about G
= 4 × 0.15 = 0.6 Nm anticlockwise

Total moment about G = 1.2 Nm anticlockwise

Notice the resultant force is zero, but the object is not in equilibrium.

Worked example

Example 2

A uniform horizontal wooden board, LM, of weight 125 N has length of 3 m and rests in equilibrium on vertical supports at A, 10 cm from L, and at B, 260 cm from L. What is the magnitude of the reaction force at each of the supports?

Solution

Draw a diagram showing the forces acting on the board and their positions.

The beam is described as uniform, so the midpoint of the beam is the centre of mass. The weight acts through that point.

The weight acts through the middle of the beam.

Let the reaction forces be RN at A and SN at B.

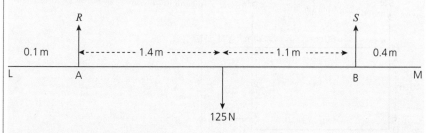

Taking moments about A

$(1.4 + 1.1) \times S - 1.4 \times 125 = 0$

$S = \dfrac{1.4 \times 125}{1.4 + 1.1} = 70$

The reaction force R has no moment about A. Taking moments here gives an equation with only one unknown S.

Resolving vertically

$R + S - 125 = 0$

Taking upwards as the positive direction.

Substituting S = 70

$R = 125 - 70 = 55$

Hence, the reaction forces are 55N at A and 70N at B.

Notice, you could also find R by taking moments about B, or any other point, should you prefer.

Worked example

Example 3

A uniform 30 cm ruler of mass 30 g is held horizontal with a downward force of 0.75 N acting on one end of the ruler and an upwards force U acting x cm from that end.

Calculate the distance x and the force U.

Solution

Draw a diagram showing the forces acting on the ruler and their positions.

The weight acts through the middle of the ruler.

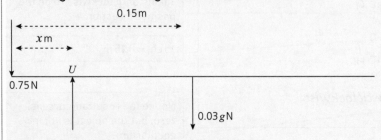

Resolving vertically $U - 0.75 - 0.03g = 0$

$$U = 1.044\,N$$

Take moments about the end of the ruler.

$1.044x - 0.03g \times 0.15 = 0$

$$x = \frac{0.03 \times 9.8 \times 0.15}{1.044} = 0.0422\,m$$

So a force of 1.044 N acts upwards 4.22 cm from the end of the ruler.

> Notice that you could take moments about the point of action of U and obtain an equation for x first if you prefer.

Worked example

Example 4

The diagram shows all the forces acting on a rectangular lamina 60 cm by 40 cm. The forces are perpendicular to the edges of the lamina which is in equilibrium. Calculate the magnitude of the forces X and Y and the distance x.

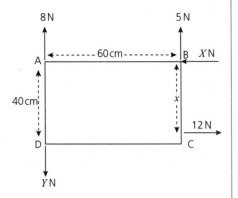

Solution

Vertically: $8 - 5 - Y = 0$ giving $Y = 3\,N$

Horizontally: $X - 12 = 0$ giving $X = 12\,N$

Take moments about A: $12x - 5 \times 0.6 = 0$

$$x = \frac{5 \times 0.6}{12} = 0.25\,m$$

> Notice that the 8 N force, the 5 N force and the forces X and Y have lines of action that pass through A and so they have no moment about that point. To find the moment of the 12 N force about A, extend the line of action of the force across the rectangle.

Worked example

Example 5

A uniform ladder of mass 6 kg and length 2.5 m stands on horizontal ground leaning against a smooth vertical wall. The bottom of the ladder is 0.8 m away from the wall.

i   Explain why the ground cannot be smooth.
ii Draw a diagram showing all the forces on the ladder.
iii Find the magnitude of the contact force between the ladder and the ground.

Solution

i The wall exerts a horizontal force on the ladder and the ladder is in equilibrium, so there must be another horizontal force, which can only be the friction between the ladder and the ground.

ii

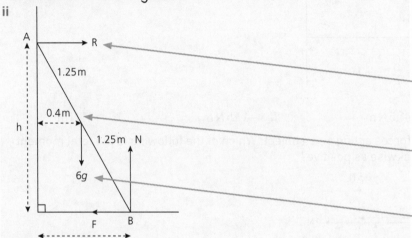

> It is useful to put the information about distances onto the force diagram.

> The wall is smooth so the contact force is at right angles to the wall.

> Notice this is half the distance the ladder is from the wall as the triangles are similar.

> The weight goes in the middle of the ladder, as the question says it is uniform.

iii Vertical equilibrium:

$N - 6g = 0$ so $N = 6g$

> Find the components of the contact force first.

To find the height of the ladder above the corner

$h = \sqrt{2.5^2 - 0.8^2} = \sqrt{5.61}$

Take moments about A:

$N \times 0.8 - 6g \times 0.4 - F \times h = 0$

> Any point can be used to take moments, but using A is easiest as the equation will not include the unknown force R.

So $F = \dfrac{1}{\sqrt{5.61}} \left(6g \times 0.8 - 6g \times 0.4 \right) = 9.93 \text{N}$

The magnitude of the contact force is

$\sqrt{N^2 + F^2} = \sqrt{(6g)^2 + 9.93^2} = 59.6 \text{N}$

> The two components F and N need to be combined together.

1 A swing is made by attaching light strings to the ends of a uniform plank of wood AB of mass 1.5 kg and length 50 cm. A girl of mass 30 kg sits in equilibrium on the swing so that her centre of mass is 20 cm from A. Find the tensions in the strings in Newtons.

A $T_A = 176.4$, $T_B = 117.6$

B $T_A = 154.35$, $T_B = 154.35$

C $T_A = 185.22$, $T_B = 123.48$

D $T_A = 183.75$, $T_B = 124.95$

2 The diagram below shows three forces acting on an object. Which of the following is the total moment about the point A, taking anticlockwise as positive?

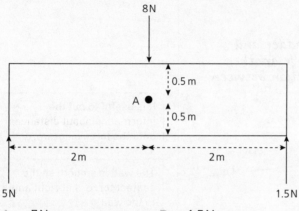

A −7 N m

B −1.5 N m

C −1.75 N m

D −11 N m

3 The diagram below shows four forces acting on an object. Which of the following is the total moment about the point P, taking anticlockwise as positive?

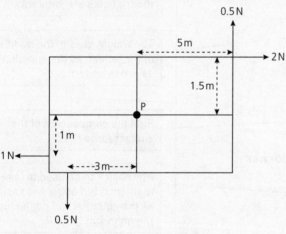

A 8 N m

B 0 N m

C 0.5 N m

D 4 N m

4 The diagram shows a light, rigid rectangle ABCD. M is the midpoint of AB.

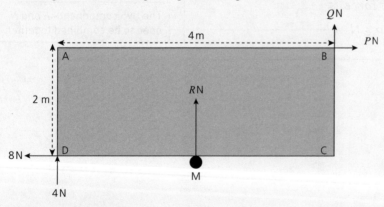

AB = 4 m and AD = 2 m.
Forces 4 N, 8 N, P N, Q N and R N act as shown.

Three of the following statements are false and one is true. Which one is true?

A   If $Q = 4$ and $P = 8$, the rectangle is in equilibrium whatever the value of R.

B   If $R = 2$, $Q = 2$ and $P = 8$, the rectangle is in equilibrium.

C   Given that the rectangle is in equilibrium, the values of P, Q and R depend on what point you take moments about.

D   Equilibrium is only possible if $P = 8$, $Q = 4$ and $R = 0$.

5   A 2 kg rectangular sign ABCD is held at a smooth hinge at A. The dimensions of the sign are shown in the diagram below. The sign is held in equilibrium with AB vertical by a horizontal force XN at B. The weight of the sign acts through its centre. Calculate the magnitude of X.

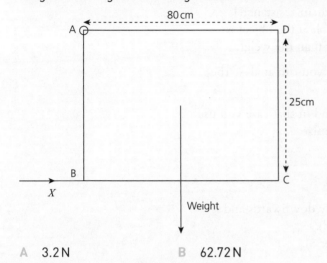

A   3.2 N   B   62.72 N   C   19.6 N   D   31.36 N

Full worked solutions online

CHECKED ANSWERS

Exam-style question

A cricket bat from an historic match has length 96 cm and mass 1.4 kg. It is displayed in a museum. It lies horizontally at rest in equilibrium on two vertical supports, which are placed 18 cm and 84 cm from the end of the handle of the bat.

In an initial model, the centre of mass, G, of the bat is taken to be at its middle.

i   Find the magnitude of the reaction force at each of the supports according to this model.

The initial model is not accurate. The supports have been carefully positioned so that the reaction force on each of them is exactly the same.

ii   Find the position of the centre of mass of the bat.

Short answers on page 226

Full worked solutions online

CHECKED ANSWERS

Chapter 13 Projectiles

About this topic

A projectile is an object moving through the air under the action of gravity. The model used here is a simple one in which air resistance is assumed to be negligible. This assumption is reasonable if the object is small and heavy, so the air resistance is much smaller than the weight.

When working with projectiles, it is important that you can analyse the horizontal and vertical motion separately.

Vector notation can also be used to help with this and in that case you use the vector versions of the constant acceleration formulae.

Before you start, remember ...

- The constant acceleration (*suvat*) equations.
- That the acceleration due to gravity, g, is vertically downwards and is usually taken to be $9.8\,\text{ms}^{-2}$ (or sometimes $10\,\text{ms}^{-2}$).
- Vertical motion under gravity (covered in Chapter 8).
- Resolving vectors (covered in Chapter 11).

Projectiles in flight

REVISED ☐

> **Key facts**
>
> 1. Projectile motion is usually considered in terms of horizontal and vertical components.
>
>   When the initial position is at the origin O:
>
>   Angle of projection = α
>
>   Initial velocity, $\mathbf{u} = \begin{pmatrix} u\cos\alpha \\ u\sin\alpha \end{pmatrix}$
>
>   Acceleration, $\mathbf{a} = \begin{pmatrix} 0 \\ -g \end{pmatrix}$
>
>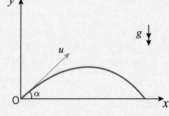
>
> 2. At time t, velocity, $\mathbf{v} = \mathbf{u} + \mathbf{a}t$ $\begin{pmatrix} v_x \\ v_y \end{pmatrix} = \begin{pmatrix} u\cos\alpha \\ u\sin\alpha \end{pmatrix} + \begin{pmatrix} 0 \\ -g \end{pmatrix}t$
>
>   Horizontal component of velocity $v_x = u\cos\alpha$
>   Vertical component of velocity $v_y = u\sin\alpha - gt$
>
> 3. At time t, position $\mathbf{r} = \mathbf{u}t + \frac{1}{2}\mathbf{a}t^2$ $\begin{pmatrix} x \\ y \end{pmatrix} = \begin{pmatrix} u\cos\alpha \\ u\sin\alpha \end{pmatrix}t + \frac{1}{2}\begin{pmatrix} 0 \\ -g \end{pmatrix}t^2$
>
>   Horizontal displacement $x = ut\cos\alpha$
>   Vertical displacement $y = ut\sin\alpha - \frac{1}{2}gt^2$
>
> 4. At the maximum height, the vertical velocity $v_y = 0$.
> 5. When the projectile lands, the vertical displacement (i.e. the height) $y = 0$
> 6. The equation of the trajectory of a projectile is found by eliminating t between the horizontal and vertical component equations. It is
>
> $$y = x\tan\alpha - \frac{gx^2}{2u^2}(1+\tan^2\alpha).$$

The examples in this section cover different aspects of the flight of a projectile which is launched from a point on level ground with initial velocity $30\,\mathrm{m\,s^{-1}}$ at an angle of $42°$ to the horizontal. The point of projection is taken to be the origin, with the x-axis horizontal and the y-axis vertical. The value of g is taken to be $9.8\,\mathrm{m\,s^{-2}}$.

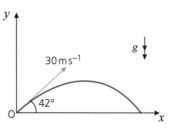

Worked example

Example 1

i   What are the initial values of the horizontal and vertical components of the velocity?

ii Write down the equations for the velocity and position after t seconds:

   A in component form

   B in vector form.

Solution

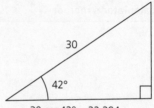

$u_y = 30 \sin 42° = 20.073\ldots$

$u_x = 30 \cos 42° = 22.294\ldots$

i   Resolving in the horizontal and vertical directions you get
$u_x = 30\cos 42° = 22.294\ldots$ and
$u_y = 30\sin 42° = 20.073\ldots$

To 3 significant figures, $u_x = 22.3$ and $u_y = 20.1$.

ii A Using $v = u + at$, $s = ut + \frac{1}{2}at^2$ in the horizontal and vertical directions:

	Velocity	Position
Horizontal	$v_x = 22.3$	$x = 22.3t$
Vertical	$v_y = 20.1 - 9.8t$	$y = 20.1t - 4.9t^2$

$u_x = 22.3$, $u_y = 20.1$
$a_x = 0$, $a_y = -9.8$

Rounded values for u_x and u_y are used as a final answer is being given.

   B Expressing these in vector form, you get

$$\begin{pmatrix} v_x \\ v_y \end{pmatrix} = \begin{pmatrix} 22.3 \\ 20.1 \end{pmatrix} + \begin{pmatrix} 0 \\ -9.8 \end{pmatrix}t \quad \text{and} \quad \begin{pmatrix} x \\ y \end{pmatrix} = \begin{pmatrix} 22.3 \\ 20.1 \end{pmatrix}t + \frac{1}{2}\begin{pmatrix} 0 \\ -9.8 \end{pmatrix}t^2$$

Worked example

Example 2

i   What is the time taken for the projectile to reach its highest point?
ii   What is the maximum height?

Solution

i   When the projectile is at its maximum height, Hm, the vertical component of the velocity is zero.

So $\qquad\qquad v_y = 0$

$\Rightarrow 20.073... - 9.8t = 0$

$\Rightarrow \qquad t = \dfrac{20.073...}{9.8} = 2.048...$

To 3 s.f. it takes 2.05 s for the projectile to reach its highest point.

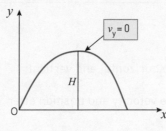

It still has a horizontal component of $22.3\,\mathrm{m\,s^{-1}}$

ii   To find the maximum height, H, you need to find y when t = 2.05.

So the maximum height, H:

$H = 2.073... \times 2.048... - 4.9 \times (2.048...)^2 = 20.559$

To 3 s.f. the height is 20.6 m.

Work with unrounded figures on your calculator and only round when you get to the final answer. You will find it helpful to store the unrounded figures you are going to use in your calculator's memory.

Worked example

Example 3

i   What is the time of flight of the projectile?
ii   What is the horizontal range?

Solution

The range, R, is the horizontal distance the projectile travels before landing, so R is the value of x when y = 0.

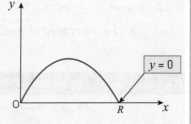

i   When y = 0, $20.073...t - 4.9t^2 = 0$

$\Rightarrow \qquad\qquad t(20.073... - 4.9t) = 0$

$\Rightarrow$ Either $\qquad\qquad\qquad t = 0$

$t = 0$ is when it leaves the ground.

or $\qquad\qquad t = \dfrac{20.073...}{4.9} = 4.096...$

$t = 4.096...$ is when it lands again.

So the time of flight is 4.10 s (3 s.f.).

Notice this is double the time used in Example 2.

ii   To find R, substitute t = 4.096... in the equation for x, so R = 22.294... × 4.096... = 91.33.

So to 3 s.f. the horizontal range is 91.3 m.

Worked example

Example 4

Show that the equation of the path of the projectile is
$y = 0.900x - 0.00986x^2$.

Solution

The path of the projectile is called its trajectory. ← It is a parabola.

You know the expressions for x and y in terms of t. They are

$$x = (22.294...)t \qquad (1)$$

and $y = (20.073...)t - 4.9t^2 \qquad (2)$

If you eliminate t, you get an equation relating x and y, which will be the equation of the trajectory.

From (1) $t = \dfrac{x}{22.294...}$

Substituting for t in (2) gives

$$y = 20.073 \times \left(\frac{x}{22.294...}\right) - 4.9 \times \left(\frac{x}{22.294...}\right)^2$$

This simplifies to give $y = 0.900x - 0.00986x^2$ *(to 3 s.f.), the equation of the trajectory.*

Eliminating t between the horizontal and vertical equations gives you the general equation of the trajectory. It is

$$y = x\tan\alpha - \frac{gx^2}{2u^2}(1 + \tan^2\alpha).$$

Test yourself

TESTED

In these questions take the upward direction as positive and use $9.8\,\mathrm{m\,s^{-2}}$ for g. All the projectiles start at the origin.

1 A particle is projected with a velocity of $30\,\mathrm{m\,s^{-1}}$ at an angle of $36.9°$ to the horizontal. What are the x- and y-coordinates of the position of the particle after time t seconds?
  A $x = 24.0t$; $y = 18.0t + 4.9t^2$
  B $x = 18.0t$; $y = 24.0t - 4.9t^2$
  C $x = 24.0t$; $y = 18.0t - 4.9t^2$
  D $x = 24.0$; $y = 18.0 - 9.8t$

2 A ball is kicked with a velocity of $14.7\,\mathrm{m\,s^{-1}}$ at an angle of $30°$ to the horizontal. Find the time taken to reach its highest point and its maximum height.
  A 1.50 s; 2.76 m
  B 0.75 s; 2.76 m
  C 0.75 s; 9.53 m
  D 1.50 s; 19.1 m

3 Find the horizontal range and time of flight of a particle that has been projected with a velocity of $21\,\mathrm{m\,s^{-1}}$ at an angle of $60°$.
  A 16.9 m; 1.86 s
  B 39.0 m; 1.86 s
  C 0 m; 3.71 s
  D 39.0 m; 3.71 s

4 What is the position, expressed in vector form, after time t seconds, of a projectile with initial velocity (in $\mathrm{m\,s^{-1}}$) of $\begin{pmatrix}12\\5\end{pmatrix}$?

  A $\begin{pmatrix}x\\y\end{pmatrix} = \begin{pmatrix}12\\5\end{pmatrix} + \begin{pmatrix}0\\-9.8\end{pmatrix}t$
  B $\begin{pmatrix}x\\y\end{pmatrix} = \begin{pmatrix}12\\5\end{pmatrix}t + \frac{1}{2}\begin{pmatrix}0\\-9.8\end{pmatrix}t^2$
  C $\begin{pmatrix}x\\y\end{pmatrix} = \begin{pmatrix}12\\5\end{pmatrix}t + \frac{1}{2}\begin{pmatrix}0\\9.8\end{pmatrix}t^2$
  D $\begin{pmatrix}x\\y\end{pmatrix} = \begin{pmatrix}5\\12\end{pmatrix} + \frac{1}{2}\begin{pmatrix}-9.8\\0\end{pmatrix}t^2$

Full worked solutions online

CHECKED ANSWERS

Exam-style question

A cricket ball is hit from ground level with a velocity of 24.5 ms⁻¹ at an angle θ to the horizontal where $\cos \theta = 0.8$ and $\sin \theta = 0.6$.

i   Show that, after t seconds, the position of the ball is given by
    $x = 19.6t$, $y = 14.7t - 4.9t^2$.

ii Find the greatest height reached by the ball.

iii Find the distance travelled before the ball bounces for the first time.

iv Show that the equation of the trajectory of the ball is $y = \dfrac{3}{4}x - \dfrac{5}{392}x^2$.

Short answers on page 226

Full worked solutions online

CHECKED ANSWERS

Further projectiles

REVISED

Key facts

1   Projectile motion is usually considered in terms of horizontal and vertical components – the convention used in this section is to take the upwards direction as positive.

2   It is important to decide at the outset where the origin O and the axes are.

3   Angle of projection $= \alpha$

    Initial velocity, $\mathbf{u} = \begin{pmatrix} u\cos\alpha \\ u\sin\alpha \end{pmatrix}$

    Initial position, $\mathbf{r_o} = \begin{pmatrix} x_o \\ y_o \end{pmatrix}$

    Acceleration, $\mathbf{a} = \begin{pmatrix} 0 \\ -g \end{pmatrix}$

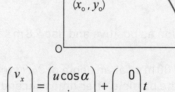

4   At time t, velocity, $\mathbf{v} = \mathbf{u} + \mathbf{a}t$ $\begin{pmatrix} v_x \\ v_y \end{pmatrix} = \begin{pmatrix} u\cos\alpha \\ u\sin\alpha \end{pmatrix} + \begin{pmatrix} 0 \\ -g \end{pmatrix} t$

    so $v_x = u\cos\alpha$
    and $v_y = u\sin\alpha - gt$

5   At time t, displacement $\mathbf{r} = \mathbf{r_o} + \mathbf{u}t + \dfrac{1}{2}\mathbf{a}t$

    $\begin{pmatrix} x \\ y \end{pmatrix} = \begin{pmatrix} x_o \\ y_o \end{pmatrix} + \begin{pmatrix} u\cos\alpha \\ u\sin\alpha \end{pmatrix} t + \dfrac{1}{2}\begin{pmatrix} 0 \\ -g \end{pmatrix} t^2$

    so $x = x_o + ut\cos\alpha$
    and $y = y_o + ut\sin\alpha - \dfrac{1}{2}gt^2$

In the following examples, the origin is not the initial position of the projectile.

Worked example

Example 1

A catapult propels a stone in an upwards direction at an angle θ, where $\cos \theta = \dfrac{7}{25}$ and $\sin \theta = \dfrac{24}{25}$, with a speed of 25 m s⁻¹ from the top of a vertical cliff which is 36 m above the sea. How long will the stone take to reach the sea? (Take $g = 10$ m s⁻².)

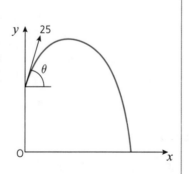

Solution

$u_x = u\cos\theta = 25 \times \frac{7}{25} = 7$ and $u_y = u\sin\theta = 25 \times \frac{24}{25} = 24$

Taking the origin as the point where the cliff meets the sea, and using

$\mathbf{r} = \mathbf{r}_0 + \mathbf{u}t + \frac{1}{2}\mathbf{a}t^2$, you get $\begin{pmatrix} x \\ y \end{pmatrix} = \begin{pmatrix} 0 \\ 36 \end{pmatrix} + \begin{pmatrix} 7 \\ 24 \end{pmatrix}t + \frac{1}{2}\begin{pmatrix} 0 \\ -10 \end{pmatrix}t^2$

and so $x = 7t$ and $y = 36 + 7t - 5t^2$

When the stone hits the water $y = 0$

that is, $y = 36 + 7t - 5t^2 = 0$, so $5t^2 - 7t - 36 = 0$

$\Rightarrow (5t + 6)(t - 6) = 0$

$\Rightarrow \qquad\qquad t = -1.2 \text{ or } 6$

As the time cannot be negative, the stone reaches the sea after 6 seconds.

Worked example

Example 2

A particle is projected from the origin. The initial components of velocity are $21\,\mathrm{m\,s^{-1}}$ horizontally and $21\,\mathrm{m\,s^{-1}}$ vertically.

i　Find expressions for x and y at time t.

ii　Eliminate t to find the equation of the trajectory.

iii　How far does the particle travel in a horizontal direction when its height is above $20\,\mathrm{m}$? (Take $g = 9.8\,\mathrm{m\,s^{-2}}$.)

Solution

i

x-direction	y-direction
$u_x = 21$, $a_x = 0$	$u_y = 21$, $a_y = -9.8$
Using $s = ut + \frac{1}{2}at^2$ in each direction	
$x = 21t$	$y = 21t - 4.9t^2$

> Some people prefer to use vector notation here.

ii　From the equation for horizontal motion $t = \frac{x}{21}$.

Substituting this value of t in the equation for vertical motion gives

$y = 21\left(\frac{x}{21}\right) - 4.9\left(\frac{x}{21}\right)^2$

This simplifies to give $y = x - \frac{x^2}{90}$. This is the equation of the trajectory.

iii　When the height is $20\,\mathrm{m}$, $y = 20$

so $20 = x - \frac{x^2}{90}$

Multiplying both sides by 90 and rearranging the terms, this equation simplifies to

$x^2 - 90x + 1800 = 0$

$\Rightarrow (x - 30)(x - 60) = 0$

so $x = 30$ or 60

So the particle will travel for $30\,\mathrm{m}$ when the height is above $20\,\mathrm{m}$ (from $x = 30$ to $x = 60$).

> Your calculator may have the facility to solve this quadratic equation.

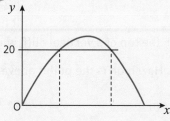

Worked example

Example 3

A particle A is projected from the top of a tower which is 12 m high with an initial velocity of $\begin{pmatrix} 15 \\ 15 \end{pmatrix}$ m s⁻¹. At the same moment another particle B is projected from the base of the tower with a velocity 30 m s⁻¹ at an angle of 60° to the horizontal. The trajectories of A and B are in the same vertical plane. Show that the particles collide 1.1 s after projection. (Take $g = 9.8\,\text{m s}^{-2}$.)

Solution

Choose the origin to be at the base of the tower.

$u_x = 30\cos 60° = 15$

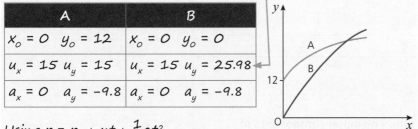

A	B
$x_0 = 0$ $y_0 = 12$	$x_0 = 0$ $y_0 = 0$
$u_x = 15$ $u_y = 15$	$u_x = 15$ $u_y = 25.98$
$a_x = 0$ $a_y = -9.8$	$a_x = 0$ $a_y = -9.8$

Using $\mathbf{r} = \mathbf{r}_0 + \mathbf{u}t + \frac{1}{2}\mathbf{a}t^2$

$$\begin{pmatrix} x_A \\ y_A \end{pmatrix} = \begin{pmatrix} 15t \\ 12 + 15t - 4.9t^2 \end{pmatrix} \text{ and } \begin{pmatrix} x_B \\ y_B \end{pmatrix} = \begin{pmatrix} 15t \\ 25.98t - 4.9t^2 \end{pmatrix}$$

If the two particles collide $x_A = x_B$ and $y_A = y_B$ for the same value of t, that is, $15t = 15t$ (which is always true) and $12 + 15t - 4.9t^2 = 25.98t - 4.9t^2$.

$\Rightarrow \qquad 12 + 15t = 25.98t$

$\Rightarrow \qquad\qquad 12 = 10.98t$

$\Rightarrow \qquad\qquad\quad t = 1.1\,\text{s (to 2 s.f.)}$

So the particles collide 1.1 s after they are projected.

> This question can be answered by using the given answer $t = 1.1$ s to find the position of each particle and show that they are at the same point. To get full marks for this method it is important to make a very clear conclusion that your working indicates that the particles collide.

Test yourself

TESTED

In these questions take the upward direction as positive and use 9.8 m s⁻² for g, except where stated otherwise.

1 A small stone is projected horizontally, at 19.6 m s⁻¹, from a height of 44.1 m above horizontal ground. What is the speed of the stone as it hits the ground?

   A $\begin{pmatrix} 19.6 \\ -29.4 \end{pmatrix}$ m s⁻¹

   B 29.4 m s⁻¹ in a downwards direction

   C 35.3 m s⁻¹

   D −9.8 m s⁻¹

2 A particle is projected from ground level with a velocity of 15 m s⁻¹ at an angle θ, where $\tan \theta = \frac{4}{3}$. For how many seconds is the particle above 7 m? (Take $g = 10\,\text{m s}^{-2}$.)

   A 1

   B 0.4

   C 1, 1.4

   D It never reaches the height of 7 m.

3 A pellet is fired from an airgun on the top of a vertical cliff, which is 12 m above sea level. The initial velocity of the pellet is $\begin{pmatrix} 7 \\ 22 \end{pmatrix}$ m s⁻¹. How high is the pellet above sea level after 4 s?

   A 28 m

   B 9.6 m

   C 21.6 m

   D 178.4 m

4 Tower A is 5.78 m high and tower B is 5.045 m high. They stand 26 m apart. From the top of tower A, a rubber bullet is fired horizontally in the direction of tower B with an initial speed of 15 m s⁻¹. A tenth of a second later a rubber bullet is fired horizontally from the top of tower B in the direction of tower A with an initial speed of 20 m s⁻¹. Which one of the following statements is true? Assume that the towers and trajectories of the bullets are all in the same vertical plane.

  A The bullets do not collide.
  B The bullets collide when the bullet fired from tower B has travelled 12 m.
  C The bullets collide at a height of 2.644 m above the ground.
  D The bullet fired from tower A reaches the ground first.

Full worked solutions online

CHECKED ANSWERS ☐

Exam-style question

Jack throws a small stone from a point that is 2 m above ground level towards a target that is 15 m away and 7.2 m above ground level. The initial velocity of the stone has horizontal and vertical components of 9 m s⁻¹ and 12 m s⁻¹, respectively. (Take $g = 9.8 \, \text{m s}^{-2}$.)

i Calculate the speed of projection, u m s⁻¹, and the angle of projection, θ, of the stone.

ii Show that, t seconds after the stone has been projected, its height above ground level, y m, is given by the expression $y = 2 + 12t - 4.9t^2$.

Find the corresponding expression for the horizontal distance, x m.

Show that the stone misses the target.

Jack tries again to hit the target, this time altering the angle of projection to 45°, but leaving the speed of projection, u m s⁻¹ unchanged.

iii What are the x- and y-coordinates of the position of the particle after t seconds?

Show that the equation of the trajectory of the stone is $y = 2 + x - \dfrac{49}{1125}x^2$. Verify that the stone now hits the target.

iv At what angle to the horizontal is the stone travelling when it hits the target?

Short answers on page 226

Full worked solutions online

CHECKED ANSWERS ☐

Chapter 14 Friction

About this topic

We live in a world full of friction, so much so that it needed a stroke of genius for Newton to realise that no force is needed to travel at constant speed.

Whenever two objects are in contact with each other there is a force between them that is normal to the contact plane, as described in Newton's 3rd law. There may also be another contact force at right angles to it. This is the friction force; it acts to oppose motion.

You need to understand the connection between these forces when an object is moving and when it is not.

Before you start, remember ...

- Drawing force diagrams.
- Resolving vectors (covered in Chapter 11).
- A particle is in equilibrium if the resultant of all the forces is zero.
- Newton's laws of motion (covered in Chapter 10).

Friction

> **Key facts**
>
> 1. The coefficient of friction between two surfaces is a measure of the roughness of the contact between the surfaces and is usually denoted by μ. If $\mu = 0$, the contact is perfectly smooth and there is no frictional force.
> 2. The frictional force F between the two surfaces is given by $F \leqslant \mu R$, where R is the normal reaction force between the two surfaces.
> 3. When the object is sliding, or on the point of sliding, $F = \mu R$ and friction is said to be limiting.
> 4. The frictional force acts to oppose motion.

Worked example

Example 1

A box of mass 10 kg is pulled along a floor by a horizontal force of 100 N. The coefficient of friction between the box and the floor is 0.8. Find the acceleration of the box.

Solution

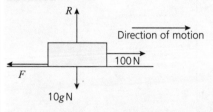

Vertically: $R - 10g = 0$ so $R = 98\,N$

Since the box is moving, friction is limiting:
$F = \mu R = 0.8 \times 98 = 78.4\,N$

Horizontally: $100 - F = ma$

$$100 - 78.4 = 10a$$

$$a = \frac{21.6}{10}$$

The acceleration is $2.16\,ms^{-2}$.

> It is a common error to mix up the resultant force and the frictional force. Newton's 2nd law is written as $F = ma$ but the F refers to the resultant force and not the friction.

Worked example

Example 2

A sledge of mass 12 kg is being pulled at constant speed along horizontal ground by a rope at 35° to the horizontal. The tension in the rope is 25 N. Calculate the coefficient of friction between the sledge and the ground.

Solution

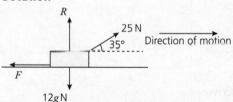

Resolving vertically: $25\sin 35° + R = 12g$

$$R = 12g - 25\sin 35°$$

Constant speed means that the forces are in equilibrium.

Resolving horizontally: $25\cos 35° - F = 0$

$$F = 25\cos 35°$$

The sledge is moving, so the friction is limiting $F = \mu R$

$F = 25\cos 35° = \mu(12g - 25\sin 35°)$

$$\mu = \frac{25\cos 35°}{12g - 25\sin 35°} = 0.198 \text{ (to 3 s.f.)}$$

> Notice the normal reaction is not equal to the weight of the sledge.

> It is often best to leave all the calculations to the end to avoid rounding errors.

The direction of the frictional force

The frictional force always opposes any tendency to motion. This is illustrated in Example 3 below. In part **i**, the block is tending to slip down the plane, and so the frictional force acts up the plane. In part **ii**, the block is moving up the plane, so the frictional force opposes this motion and acts down the plane.

Worked example

Example 3

A block of mass 2 kg rests on a plane which is inclined at 20° to the horizontal. The coefficient of friction between the plane and the block is 0.2. The block is held in place by a string parallel to the plane.

i   What is the minimum tension in the string needed to hold the block in place?

ii What is the tension in the string if the block is on the point of moving up the plane?

Solution

Let the tension in the string be T.

i   *Perpendicular to the plane:*

$$R - 2g\cos 20° = 0$$
$$R = 19.6\cos 20°$$

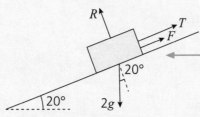

> When the minimum tension is applied, the block is about to slide down the plane, so the friction is limiting and acts up the plane.

Friction is limiting: $F = \mu R = 0.2 \times 19.6\cos 20°$
$$= 3.92\cos 20°$$

Parallel to the plane: $T + F - 2g\sin 20° = 0$
$$T = 19.6\sin 20° - 3.92\cos 20° = 3.02$$

The minimum tension required is 3.02 N (3 s.f.).

ii *There is no change in the forces perpendicular to the plane, so the magnitude of the frictional force is the same as in part i.*

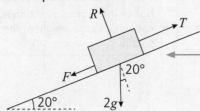

> When the block is about to slide up the plane, friction is limiting and acts down the plane.

Parallel to the plane: $T - F - 2g\sin 20° = 0$
$$T = 19.6\sin 20° + 3.92\cos 20° = 10.4$$

The tension is 10.4 N (3 s.f.).

Test yourself

1 Hakim pushes a box of mass 12 kg across a rough horizontal floor. The coefficient of friction between the box and the floor is 0.5. He applies a horizontal force of 75 N. Calculate the acceleration of the box.

   A $5.75\,\text{m}\,\text{s}^{-2}$ B $1.35\,\text{m}\,\text{s}^{-2}$ C $0.138\,\text{m}\,\text{s}^{-2}$ D $3.125\,\text{m}\,\text{s}^{-2}$

2 A block of mass 2 kg rests on a rough slope which makes an angle of α with the horizontal. It is on the point of sliding down the slope. What is the coefficient of friction between the block and the slope?

   A $\sin\alpha$ B $\cos\alpha$ C $\tan\alpha$ D $\dfrac{1}{\cos\alpha}$

3 A block of mass 5 kg is to be pulled along a rough horizontal floor by a rope which is inclined at 40° to the horizontal. The coefficient of friction between the block and the floor is 0.6. What is the tension in the rope when the block is on the point of moving?

   A 25.5 N   B 32.4 N   C 38.4 N   D 2.6 N

4 A particle is placed on a rough horizontal plane inclined at 50° to the horizontal. The coefficient of friction between the particle and the plane is 0.4. The particle is pulled up the slope with acceleration $2\,\text{m}\,\text{s}^{-2}$ by a constant force $P\,\text{N}$ which acts parallel to the plane. What is the value of P?

   A 14.0 N   B 20.1 N   C 24.1 N   D 22.6 N

5 A block of mass 0.5 kg moves upwards along a slope which makes an angle of 15° with the horizontal. Its acceleration is $-4\,\text{m}\,\text{s}^{-2}$. Calculate the coefficient of friction between the block and the plane, giving your answer correct to 3 significant figures.

   A 0.154   B 0.149   C 0.423   D 0.155

Full worked solutions online

Exam-style question

A 3 kg block is placed on a rough plane inclined at 40° to the horizontal. The coefficient of friction between the block and the plane is 0.4. A light inextensible string which is parallel to the slope is attached to the block. The string passes over a smooth pulley and is attached to a particle of mass 5 kg which hangs freely. The system is released from rest and the block is pulled up the slope.

i Calculate the acceleration of the system.

When the block is travelling at $1.5\,\text{m}\,\text{s}^{-1}$, the string snaps.

ii Calculate the distance that the block travels before coming to rest.

Short answers on page 226

Full worked solutions online

Review questions (Mechanics)

1 A boy starts at the origin and jogs to A which is 60 m due North in 8 s. He stays at A for 18 s and then jogs 30 m due South to B at the same speed as before. Find:

   i   the displacement after both parts of the journey

   ii the total distance travelled

   iii the total time taken

   iv the average speed.

2 The motion of a particle is illustrated by the velocity–time graph. The positive direction is due North and the particle begins 200 m north of the origin.

   i   Describe the motion of the particle.

   ii State the acceleration for each phase of the motion.

   iii Find the position of the particle after 40 s.

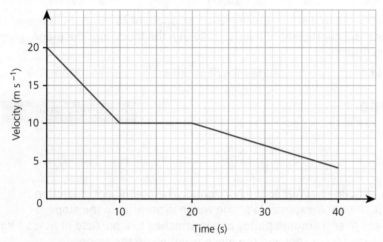

3 A space probe travels in a straight line. It accelerates uniformly from an initial velocity of 24 m s^{-1} in the negative direction to a velocity of 120 m s^{-1} in the positive direction in 120 s. It then continues to move along the same straight line with the same acceleration.

   Find how long it takes the space probe to reach a point 50 km from its starting position. Give your answer in minutes and seconds to the nearest second.

4 Anastasia throws a ball vertically upwards from a point 1.2 m above the ground in her garden with a velocity of 15 m s^{-1}. The garden is surrounded by a hedge that is 3 m high.
   For how long is the ball visible from outside the garden?

5 In this question all quantities are in S.I. units. The acceleration of a particle is given by the vector $\mathbf{a} = \mathbf{i} - 2\mathbf{j}$, where $\mathbf{i}$ and $\mathbf{j}$ are unit vectors East and North, respectively. At time $t = 0$, the particle has a velocity $\mathbf{u} = 6\mathbf{i} - 2\mathbf{j}$ and position vector $\mathbf{r}_0 = -14\mathbf{i} + 8\mathbf{j}$.

   i   Show that the particle reaches the origin when $t = 2$ s.

   ii Calculate the speed of the particle when $t = 2$ s.

6 Jenny stands on the end of a pier 4 m above the sea. She throws a pebble at 12 m s^{-1} at 35° above the horizontal out to sea.

   i   Calculate the maximum height of the pebble above the point of projection.

   ii Calculate the horizontal distance travelled by the pebble when it hits the sea.

   iii Find the direction of motion of the particle as it hits the sea.

7 A toy train consists of an engine of mass 8 kg and a carriage of mass 6 kg. It is travelling on level ground along a straight track at constant speed. The resistances to motion on the engine and the carriage are 2 N and 1.5 N, respectively. The normal reactions between the engine and the track and the carriage and the track are N_1 and N_2. The engine exerts a driving force of D N. The tension in the coupling is T N.

   i Draw a diagram showing all the forces acting on the engine and on the carriage.

   ii Find the values of D, N_1, N_2 and T.

8 A woman of mass 56 kg is in a lift which is ascending. Find the contact force between the woman and the lift floor in each case.

   i The lift is accelerating upwards at $2\,\mathrm{m\,s^{-2}}$.

   ii The lift is travelling at constant speed.

   iii The lift is slowing down with a deceleration of $3\,\mathrm{m\,s^{-2}}$.

9 A car of mass 1500 kg pulls a caravan of mass 1200 kg. It accelerates from rest to $12\,\mathrm{m\,s^{-1}}$ in 15 seconds.

   i Calculate the acceleration of the car and caravan assuming that it is constant.

   In an initial model, it is assumed that there is no resistance to motion.

   ii Find the driving force the car must provide according to this model.

   In a refined model it is assumed that the resistance is not negligible and that the driving force is 2500 N.

   iii Calculate the total resistance to motion according to this model.

   In this model it is also assumed that the resistance on the caravan is four times the resistance on the car.

   iv Calculate the tension in the tow-bar.

10 In this question, the x and y directions are East and North, respectively.

   Forces X and Y act on a particle of mass 2 kg. X has a magnitude of 5 N and acts along a bearing of 060°. Y has a magnitude of 4 N and acts due West.

   i Find the resultant of X and Y in vector form.

   ii Find the acceleration of the particle.

   iii The initial velocity of the particle is $\begin{pmatrix} 5 \\ -6 \end{pmatrix}\,\mathrm{m\,s^{-1}}$. Find the velocity after 5 seconds.

11 A block of mass 8 kg rests on a smooth plane inclined at 23° to the horizontal. It is held in place by a string which makes an angle of 10° to the plane as shown in the diagram.

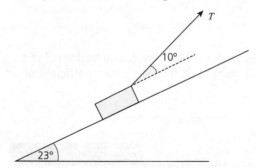

   i Find the tension in the string.

   ii Find the normal reaction force between the block and the plane.

   The tension is increased to 40 N.

   iii Find the acceleration of the block up the plane.

12 Sven uses a rope to pull a sledge of mass 45 kg across a rough horizontal surface. The rope makes an angle of 35° with the ground. The acceleration of the sledge is 2.5 m s⁻² and the tension in the rope is 215 N.

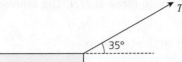

   i Calculate the magnitude of the frictional force acting on the sledge.

   ii Calculate the coefficient of friction between the sledge and the surface, giving your answer correct to 3 significant figures.

13 The diagram shows a rectangular lamina ABCD where the distances AB and BC are 0.3 m and 0.5 m, respectively. Forces of 30 N and 40 N act at A and B at right angles to the edge AB. A force F is applied to the edge CD to keep the lamina in equilibrium.

   i Find the magnitude of the force F and the distance from D of the point of application.

   The force F is now applied to M which is the midpoint of the edge CD. Additional forces at right angles to the lines AD and BC are applied at A and C to maintain equilibrium.

   ii Find these forces, indicating clearly the direction in which they act.

14 A boy starts running the 100 m race from rest at the origin. His velocity v m s⁻¹ at time t seconds is given by

   $v = 4.2t - 0.6t^2$ for $0 \leqslant t \leqslant 3.5$

   $v = 7.35$ from $t = 3.5$ until the end of the race.

   i Find an expression for his acceleration at time t s.
   Show that his acceleration is zero when $t = 3.5$, and interpret this result.

   ii Find the distance he has run after:

   A 3.5 s

   B 9.5 s.

   iii Calculate the time he takes to run 100 m.

15 The velocity in m s⁻¹ of a particle at time t seconds is given by $\mathbf{v} = \begin{pmatrix} 5 \\ 3t^2 - 1 \end{pmatrix}$. The initial position of the particle is $\mathbf{r} = \begin{pmatrix} 10 \\ -4 \end{pmatrix}$ m from the origin.

   i Find an expression for the acceleration of the particle at time t and determine whether the acceleration is constant.

   ii Find the position of the particle after 3 seconds.

   iii Find the cartesian equation of the path of the particle.

16 A uniform ladder of mass 4 kg which is 3 m long leans against a smooth wall. The bottom of the ladder is 1 m away from the wall and the ladder is on the point of sliding. Find the coefficient of friction between the ladder and the floor.

Short answers on page 226

Full worked solutions online

CHECKED ANSWERS

Exam preparation

Before your exam

- *Start revising early* – half an hour a day for 6 months is better than cramming in all–nighters in the week before the exam. Little and often is the key.
- *Don't procrastinate* – you won't feel more like revising tomorrow than you do today!
- Put your phone on *silent* while you revise – don't get distracted by a constant stream of messages from your friends.
- Make sure your *notes are in order* and nothing is missing.
- *Be productive* – don't waste time colouring endless revision timetables. Make sure your study time is actually spent revising!
- Use the 'Target Your Revision' sections to *focus your revision* on the topics you find tricky – remember you won't improve if you only answer the questions you could do anyway!
- Don't just read about a topic. *Maths is an active subject* – you improve by answering questions and actually *doing* maths, not just reading about it.
- Cover up the solution to an example and then try to answer it yourself.
- Answer as many past exam questions as you can. Work through the *'Review'* questions first and then move on to past papers.
- Try *teaching a friend* a topic. Teaching something is the best way to learn it yourself – that's why your teachers know so much!

The exam papers

You must take **all** of Components 01, 02 and 03 to be awarded the OCR A Level in Mathematics (A).

Component	Title	No. of marks		
01	Pure Mathematics	100	2 hours Written paper	$33\frac{1}{3}$% of total A Level
02	Pure Mathematics and Statistics	100	2 hours Written paper	$33\frac{1}{3}$% of total A Level
03	Pure Mathematics and Mechanics	100	2 hours Written paper	$33\frac{1}{3}$% of total A Level

The content of this book covers all the Statistics content for component 02 and the Mechanics content for component 03.

Make sure you know these formulae for your exam ...

From GCSE Maths you need to know ...

Topic	Formula
Circle	$Area = \pi r^2$ $Circumference = 2\pi r$, where r is the radius
Parallelogram 	$Area = base \times vertical\ height$
Trapezium 	$Area = \frac{1}{2}h(a+b)$
Triangle 	$Area = \frac{1}{2}\ base \times vertical\ height$
Prism	$Volume = area\ of\ cross\ section \times length$
Cylinder	$Volume = \pi r^2 h$ $Area\ of\ curved\ surface = 2\pi rh$ $Total\ surface\ area = 2\pi rh + 2\pi r^2$, where r is the radius and h is the height
Pythagoras' theorem 	$a^2 + b^2 = c^2$
Trigonometry 	$\cos\theta = \dfrac{adjacent}{hypotenuse}$ $\sin\theta = \dfrac{opposite}{hypotenuse}$ $\tan\theta = \dfrac{opposite}{adjacent}$

Topic	Formula	
Circle theorems	The angle in a semi-circle is a right-angle.	
	The perpendicular from the centre of a circle to a chord bisects the chord.	
	The tangent to a circle at a point is perpendicular to the radius through that point.	
Speed	$speed = \dfrac{distance}{time}$	
Distance–time graphs	$gradient = speed$	
Velocity–time graphs	$gradient = acceleration$	
	$area = displacement$	

From A Level Pure Maths you should know...

Topic	Formula
Laws of indices	$a^m \times a^n = a^{m+n}$
	$\dfrac{a^m}{a^n} = a^{m-n}$
	$(a^m)^n = a^{mn}$
	$a^{-n} = \dfrac{1}{a^n}$
	$\sqrt[n]{a} = a^{\frac{1}{n}}$
	$\sqrt[n]{a^m} = a^{\frac{m}{n}}$
	$a^0 = 1$
Quadratic equations	The quadratic equation $ax^2 + bx + c = 0$ has roots $x = \dfrac{-b \pm \sqrt{b^2 - 4ac}}{2a}$

Topic	Formula
Coordinate geometry	• For two points (x_1, y_1) and (x_2, y_2): $\text{Gradient} = \dfrac{y_2 - y_1}{x_2 - x_1}$ $\text{Length} = \sqrt{(x_2 - x_1)^2 + (y_2 - y_1)^2}$ $\text{Midpoint} = \left(\dfrac{x_1 + x_2}{2}, \dfrac{y_1 + y_2}{2} \right)$ • The equation of a straight line with gradient m and y-intercept $(0, c)$ is $y = mx + c$. • The equation of a straight line with gradient m and passing through (x_1, y_1) is $y - y_1 = m(x - x_1)$. • Parallel lines have the same gradient. • For two perpendicular lines, $m_1 m_2 = -1$. • The equation of a circle, centre (a, b) and radius r is $(x - a)^2 + (y - b)^2 = r^2$.
Trigonometry	For any triangle ABC: $\text{Area} = \dfrac{1}{2} ab \sin C$ $\text{Sine rule: } \dfrac{a}{\sin A} = \dfrac{b}{\sin B} = \dfrac{c}{\sin C}$ or $\quad \dfrac{\sin A}{a} = \dfrac{\sin B}{b} = \dfrac{\sin C}{c}$ Cosine Rule: $a^2 = b^2 + c^2 - 2bc \cos A$ or $\quad \cos A = \dfrac{b^2 + c^2 - a^2}{2bc}$ Identities: $\sin^2 \theta + \cos^2 \theta \equiv 1$ $\tan \theta \equiv \dfrac{\sin \theta}{\cos \theta}, \quad \cos \theta \neq 0$ $\sec^2 A \equiv 1 + \tan^2 A$ $\operatorname{cosec}^2 A \equiv 1 + \cot^2 A$ $\sin 2A \equiv 2 \sin A \cos A$ $\cos 2A \equiv \cos^2 A - \sin^2 A$ $\tan 2A \equiv \dfrac{2 \tan A}{1 - \tan^2 A}$ For a circle of radius r, where an angle at the centre of θ radians subtends an arc of length s and encloses an associated sector of area A: $s = r\theta \quad A = \dfrac{1}{2} r^2 \theta$
Transformations	$y = \text{f}(x + a)$ is a translation of $y = \text{f}(x)$ by $\begin{pmatrix} -a \\ 0 \end{pmatrix}$ $y = \text{f}(x) + b$ is a translation of $y = \text{f}(x)$ by $\begin{pmatrix} 0 \\ b \end{pmatrix}$ $y = \text{f}(ax)$ is a one–way stretch of $y = \text{f}(x)$, parallel to x-axis, scale factor $\dfrac{1}{a}$ $y = a\text{f}(x)$ is a one–way stretch of $y = \text{f}(x)$, parallel to y-axis, scale factor a $y = \text{f}(-x)$ is a reflection of $y = \text{f}(x)$ in the y-axis $y = -\text{f}(x)$ is a reflection of $y = \text{f}(x)$ in the x-axis

Topic	Formula
Polynomials and binomial expansions	*The Factor theorem*:
	If $(x - a)$ is a factor of f(x) then f$(a) = 0$ and $x = a$ is a root of the equation f$(a) = 0$.
	Conversely, if f$(a) = 0$ then $(x - a)$ is a factor of f(x).
	Pascal's triangle:
	$$\begin{array}{ccccccccc} & & & & 1 & & & & \\ & & & 1 & & 1 & & & \\ & & 1 & & 2 & & 1 & & \\ & 1 & & 3 & & 3 & & 1 & \\ 1 & & 4 & & 6 & & 4 & & 1 \end{array}$$
	Factorials: $n! = n \times (n-1) \times (n-2) \times \dots \times 1$

Differentiation	*Function*	*Derivative*
	$y = kx^n$	$\dfrac{dy}{dx} = knx^{n-1}$
	$y = e^{kx}$	$\dfrac{dy}{dx} = ke^{kx}$
	$y = f(x) + g(x)$	$\dfrac{dy}{dx} = f'(x) + g'(x)$
	$y = \sin kx$	$\dfrac{dy}{dx} = k\cos kx$
	$y = \cos kx$	$\dfrac{dy}{dx} = -k\sin kx$
	$y = \ln x$	$\dfrac{dy}{dx} = \dfrac{1}{x}$

The Chain rule for differentiating a function of a function is $\dfrac{dy}{dx} = \dfrac{dy}{du} \times \dfrac{du}{dx}$.

The Product rule: $y = uv$, $\dfrac{dy}{dx} = u\dfrac{dv}{dx} + v\dfrac{du}{dx}$.

Integration	*Function*	*Integral*
	$\displaystyle\int kx^n \, dx$	$\dfrac{kx^{n+1}}{n+1} + c$
	$\cos kx$	$\dfrac{1}{k}\sin kx + c$
	$\sin kx$	$-\dfrac{1}{k}\cos kx + c$
	e^{kx}	$\dfrac{1}{k}e^{kx} + c$
	$\dfrac{1}{x}$	$\ln|x| + c, \; x \neq 0$
	$f'(x) + g'(x)$	$f(x) + g(x) + c$
	$f'(g(x))g'(x)$	$f(g(x)) + c$

Area under a curve $= \displaystyle\int_a^b y \, dx \; (y \geqslant 0)$

Vectors	$\overrightarrow{AB} = \overrightarrow{OB} - \overrightarrow{OA}$
	If $\mathbf{a} = x\mathbf{i} + y\mathbf{j}$ then $|\mathbf{a}| = \sqrt{x^2 + y^2}$

Topic	Formula
Exponentials and logarithms	$y = \log_a x \Leftrightarrow a^y = x$ for $a \geqslant 0$ and $x \geqslant 0$
	$\log xy = \log x + \log y \qquad \log \sqrt[n]{x} = \dfrac{1}{n}\log x \quad \log_a a = 1$
	$\log \dfrac{x}{y} = \log x - \log y \qquad \log \dfrac{1}{x} = -\log x$
	$\log x^n = n\log x \qquad\qquad \log 1 = 0 \qquad\quad \log_e x = \ln x \qquad\quad e = 2.718\ldots$
Sequences and series	*Arithmetic sequence:*
	The kth term is $a_k = a + (k-1)d$
	The last term, $l = a_n = a + (n-1)d$
	Geometric sequence:
	The kth term is $a_k = ar^{k-1}$
	The last term, $a_n = ar^{n-1}$

From A Level Mechanics you should know ...

Topic	Formula
Forces and equilibrium	Weight $= mg$
	Newton's second law in the form $F = ma$
Vector form of *suvat* equations	$\mathbf{v} = \mathbf{u} + \mathbf{a}t$
	$\mathbf{r} = \mathbf{u}t + \dfrac{1}{2}\mathbf{a}t^2$
	$\mathbf{r} = \dfrac{1}{2}(\mathbf{u} + \mathbf{v})$
	$\mathbf{r} = \mathbf{v}t - \dfrac{1}{2}\mathbf{a}t^2$
Calculus expressions	displacement $\qquad\qquad$ velocity $\qquad\qquad$ acceleration
	Differentiate $\longrightarrow$
	$s \qquad\qquad v = \dfrac{ds}{dt} \qquad\qquad a = \dfrac{dv}{dt} = \dfrac{d^2 s}{dt^2}$
	$\longleftarrow$ *Integrate*
	$s = \displaystyle\int v\,dt \qquad\qquad v = \displaystyle\int a\,dt \qquad\qquad a$
Vector forms of calculus expressions	displacement $\qquad\qquad$ velocity $\qquad\qquad$ acceleration
	Differentiate $\longrightarrow$
	$\mathbf{r} = x\mathbf{i} + y\mathbf{j} \qquad \mathbf{v} = \dfrac{d\mathbf{r}}{dt} = \dot{x}\mathbf{i} + \dot{y}\mathbf{j} \qquad \mathbf{a} = \dfrac{d\mathbf{v}}{dt} = \dfrac{d^2\mathbf{r}}{dt^2} = \ddot{x}\mathbf{i} + \ddot{y}\mathbf{j}$
	$\mathbf{r} = \begin{pmatrix} x \\ y \end{pmatrix} \qquad\qquad \mathbf{v} = \begin{pmatrix} \dot{x} \\ \dot{y} \end{pmatrix} \qquad\qquad \mathbf{a} = \begin{pmatrix} \ddot{x} \\ \ddot{y} \end{pmatrix}$
	$\longleftarrow$ *Integrate*
	$\mathbf{r} = \displaystyle\int \mathbf{v}\,dt \qquad\qquad \mathbf{v} = \displaystyle\int \mathbf{a}\,dt \qquad\qquad \mathbf{a}$
Scalar quantities	*Distance* $r = |\mathbf{r}| = \sqrt{x^2 + y^2}$
	Speed $v = |\mathbf{v}| = \sqrt{\dot{x}^2 + \dot{y}^2}$
	Magnitude of acceleration $a = |\mathbf{a}| = \sqrt{\ddot{x}^2 + \ddot{y}^2}$

Topic	Formula
Vector form of Newton's 2nd law	$\mathbf{F} = m\mathbf{a}$
Resolving a force	$\mathbf{F} = F\cos\theta\,\mathbf{i} + F\sin\theta\,\mathbf{j}$
Moment of a force	Moment $= Fd$
Projectiles	$\begin{pmatrix} v_x \\ v_y \end{pmatrix} = \begin{pmatrix} u\cos\alpha \\ u\sin\alpha \end{pmatrix} + \begin{pmatrix} 0 \\ -g \end{pmatrix} t$ $y = \begin{pmatrix} u\cos\alpha \\ u\sin\alpha \end{pmatrix} t + \frac{1}{2}\begin{pmatrix} 0 \\ -g \end{pmatrix} t^2$ Equation of trajectory: $y = x\tan\alpha - \dfrac{gx^2}{2u^2\cos^2\alpha} = x\tan\alpha - \dfrac{gx^2\left(1+\tan^2\alpha\right)}{2u^2}$
Friction	$F \leqslant \mu R$

From A Level Statistics you should know ...

Topic	Formula
The mean of a set of data	$\bar{x} = \dfrac{\sum x}{n} = \dfrac{\sum fx}{\sum f}$
The standard Normal variable	$Z = \dfrac{X-\mu}{\sigma}$ where $X \sim N(\mu,\ \sigma^2)$

Formulae that will be given

Make sure you are familiar with the formulae book you will use in the exam.

The formulae sheet is subject to change so always check the latest version on the exam board's website.

Here are the formulae you are given for the exam.

Topic	Formula
Arithmetic series	$S_n = \frac{1}{2}n(a+l) = \frac{1}{2}n\{2a+(n-1)d\}$
Geometric series	$S_n = \dfrac{a(1-r^n)}{1-r}$ $S_\infty = \dfrac{a}{1-r}$ for $\|r\| < 1$
Binomial series	$(a+b)^n = a^n + {}^nC_1 a^{n-1}b + {}^nC_2 a^{n-2}b^2 + \ldots + {}^nC_r a^{n-r}b^r + \ldots + b^n \ (n \in \mathbb{N})$, where ${}^nC_r = {}_nC_r = \begin{pmatrix} n \\ r \end{pmatrix} = \dfrac{n!}{r!(n-r)!}$ $(1+x)^n = 1 + nx + \dfrac{n(n-1)}{2!}x^2 + \ldots + \dfrac{n(n-1)\ldots(n-r+1)}{r!}x^r + \ldots \quad (\|x\| < 1, n \in \mathbb{R})$
Differentiation	<table><tr><td>$f(x)$</td><td>$f'(x)$</td></tr><tr><td>$\tan kx$</td><td>$k\sec^2 kx$</td></tr><tr><td>$\sec x$</td><td>$\sec x \tan x$</td></tr><tr><td>$\cot x$</td><td>$-\csc^2 x$</td></tr><tr><td>$\csc x$</td><td>$-\csc x \cot x$</td></tr></table> Quotient rule: $y = \dfrac{u}{v}, \quad \dfrac{dy}{dx} = \dfrac{v\frac{du}{dx} - u\frac{dv}{dx}}{v^2}$ Differentiation from first principles: $f'(x) = \lim\limits_{h \to 0} \dfrac{f(x+h) - f(x)}{h}$
Integration	$\displaystyle\int \dfrac{f'(x)}{f(x)}\,dx = \ln\|f(x)\| + c$ $\displaystyle\int f'(x)\big(f(x)\big)^n\,dx = \dfrac{1}{n+1}\big(f(x)\big)^{n+1} + c$ Integration by parts: $\displaystyle\int u\dfrac{dv}{dx}\,dx = uv - \int v\dfrac{du}{dx}\,dx$
Small angle approximations	$\sin\theta \approx \theta$, $\cos\theta \approx 1 - \frac{1}{2}\theta^2$, $\tan\theta \approx \theta$ where θ is measured in radians.
Trigonometric identities	$\sin(A \pm B) = \sin A \cos B \pm \cos A \sin B$ $\cos(A \pm B) = \cos A \cos B \mp \sin A \sin B$ $\tan(A \pm B) = \dfrac{\tan A \pm \tan B}{1 \mp \tan A \tan B} \qquad (A \pm B \neq (k + \frac{1}{2})\pi)$
Numerical methods	Trapezium rule: $\displaystyle\int_a^b y\,dx \approx \frac{1}{2}h\{(y_0 + y_n) + 2(y_1 + y_2 + \ldots + y_{n-1})\}$, where $h = \dfrac{b-a}{n}$ The Newton–Raphson iteration for solving $f(x) = 0$: $x_{n+1} = x_n - \dfrac{f(x_n)}{f'(x_n)}$
Probability	$P(A \cup B) = P(A) + P(B) - P(A \cap B)$ $P(A \cap B) = P(A)P(B\|A) = P(B)P(A\|B)$ or $P(A\|B) = \dfrac{P(A \cap B)}{P(B)}$

Topic	Formula
Standard deviation	standard deviation $= \sqrt{\dfrac{\sum (x - \bar{x})^2}{n}} = \sqrt{\dfrac{\sum x^2}{n} - \bar{x}^2}$ or $\sqrt{\dfrac{\sum f(x - \bar{x})^2}{\sum f}} = \sqrt{\dfrac{\sum fx^2}{\sum f} - \bar{x}^2}$
The binomial distribution	If $X \sim B(n, p)$ then $P(X = r) = \dbinom{n}{x} p^x (1 - p)^{n-x}$ Mean of X is np, Variance of X is $np(1 - p)$
Hypothesis testing for the mean of a Normal distribution	If $X \sim N(\mu, \sigma^2)$ then $\bar{X} \sim N\left(\mu, \dfrac{\sigma^2}{n}\right)$ and $\dfrac{\bar{X} - \mu}{\sigma / \sqrt{n}} \sim N(0,1)$
Percentage points of the Normal distribution	If Z has a normal distribution with mean 0 and variance 1 then, for each value of p, the table gives the value of z such that $P(Z \leq z) = p$.

p	0.75	0.90	0.95	0.975	0.99	0.995	0.9975	0.999	0.9995
z	0.674	1.282	1.645	1.960	2.326	2.575	2.807	3.090	3.291

Kinematics

Motion in a straight line:

$v = u + at$

$s = ut + \frac{1}{2}at^2$

$s = \frac{1}{2}(u + v)t$

$v^2 = u^2 + 2as$

$s = vt - \frac{1}{2}at^2$

Motion in two dimensions:

$\mathbf{v} = \mathbf{u} + \mathbf{a}t$

$\mathbf{s} = \mathbf{u}t + \frac{1}{2}\mathbf{a}t^2$

$\mathbf{s} = \frac{1}{2}(\mathbf{u} + \mathbf{v})t$

$\mathbf{s} = \mathbf{v}t - \frac{1}{2}\mathbf{a}t^2$

Make sure you know these formulae for your exam …

During your exam

Watch out for these key words:

- **Exact ...** leave your answer as a simplified surd, fraction or power.

  Examples: $\ln 5$ ✓ 1.61 ✗ e^2 ✓ 7.39 ✗

  $2\sqrt{3}$ ✓ 3.26 ✗ $1\frac{5}{6}$ ✓ 1.83 ✗

- **Determine ...** justification should be given for any results found, including working where appropriate.

- **Give/State/Write down ...** no working is expected – unless it helps you. The marks are for the answer rather than the method.

  Example: The equation of a circle is $(x+2)^2 + (y-3)^2 = 13$. Write down the radius of the circle and the coordinates of its centre. ◄──── Make sure you give both answers!

- **Prove/Show that ...** the answer has been given to you. You must show full working otherwise you will lose marks. Often you will need the answer to this part to answer the next part of the question. Most of the marks will be for the method.

  Example: i Prove that $\sin x - \cos^2 x \equiv \sin^2 x + \sin x - 1$.

- **Hence ...** you must follow on from the given statement or previous part. Alternative methods may not earn marks.

  Example: ii Hence solve $\sin x - \cos^2 x = -2$ for $0° \leqslant x \leqslant 180°$. ◄──── Remember that if you couldn't answer part **i** you can still go on and answer part **ii**.

- **Hence or otherwise ...** there may be several ways you can answer this question but it is likely that following on from the previous result will be the most efficient and straightforward method.

  Example: Factorise $p(x) = 6x^2 + x - 2$.

  Hence, or otherwise, solve $p(x) = 0$.

- **You may use the result ...** indicates a given result that you would not always be expected to know, but which may be useful in answering the question. This does not necessarily mean that you have to use the given result.

- **Plot ...** you should mark points accurately on graph paper provided in the Printed Answer Booklet. You will either have been given the points or have had to calculate them. You may also need to join them with a curve or a straight line.

  Example: Plot this additional point on the scatter diagram.

- **Draw ...** you should draw to an accuracy appropriate to the problem. You are being asked to make a sensible judgement about the level of accuracy which is appropriate.

  Example: Draw a diagram showing the forces acting on the particle. Draw a line of best fit for the data.

- **Sketch (a graph) ...** you should draw a diagram, not necessarily to scale, showing the main features of a curve. These are likely to include at least some of the following:
  - Turning points
  - Asymptotes
  - Intersection with the y-axis
  - Intersection with the x-axis
  - Behaviour for large y (+ or −)

  Any other important features should also be shown.

  Example: Sketch the curve with equation $y = \dfrac{1}{(x-1)}$.

- **In this question you must show detailed reasoning** … you must give a solution which leads to a conclusion showing a detailed and complete analytical method. Your solution should contain sufficient detail to allow the line of your argument to be followed. This does not prevent you from using a calculator when tackling the question, e.g. for checking an answer or evaluating a function at a given point, but it is a restriction on what will be accepted as evidence of a complete method.

Watch out for these common mistakes:

- ✗ Miscopying your own work or misreading/miscopying the question.
- ✗ Not giving your answer as coordinates when it should be, e.g. when finding where two curves meet.
- ✗ Check the units in the question, e.g. length given in cm not m.
- ✗ Giving your answer as coordinates inappropriately, e.g. writing a vector as coordinates.
- ✗ Not finding y-coordinates when asked to find coordinates.
- ✗ Not finding where the curve cuts the x- **and** y-axes when sketching a curve.
- ✗ Using a ruler to draw curves.
- ✗ Not using a ruler to draw straight lines.
- ✗ Spending too long drawing graphs when a sketch will do.
- ✗ Not stating the equations of asymptotes when sketching an exponential or reciprocal curve.
- ✗ Not simplifying your answer sufficiently.
- ✗ Rounding answers that should be **exact**.
- ✗ Rounding errors – don't round until you reach your final answer.
- ✗ Giving answers to the wrong degree of accuracy – use 3 s.f. unless the questions says otherwise.
- ✗ Not showing any or not showing enough working – especially in 'show that' or 'proof' questions.
- ✗ Showing full working for a mean or standard deviation calculation when you were expected to write down an answer using your calculator functions.
- ✗ Being too assertive when giving the final conclusion to a statistical hypothesis test.
- ✗ Not saying what the variables you introduce stand for.
- ✗ Giving two contradictory answers – you need to make it clear which your answer is if you change your mind.

Once you have answered a question **re–read the question** making sure you've answered **all** of it. It is easy to miss the last little bit of a question.

Check your answer is …

- ✓ to the correct accuracy
- ✓ a sensible size in the context of the question
- ✓ in the right form
- ✓ complete…have you answered the whole question?

If you do get stuck …

Keep calm and don't panic.
- ✓ **Reread the question** … have you skipped over a key piece of information that would help? Highlight any numbers or key words.
- ✓ **Draw** … a diagram often helps. Especially in questions on graphs, coordinate geometry and vectors, a sketch can help you see the way forward.

- ✓ **Look** … for how you can re−enter the question. Not being able to answer part i doesn't mean you won't be able to do part ii. Remember the last part of a question is not necessarily harder.
- ✓ **Mark allocation** … look at how many marks are allocated to the part of the questions you are stuck on. If it is only one or two, and you are doing lots of work, you are probably wasting your time going in a wrong direction.
- ✓ **Move on** … move onto the next question or part question. Don't waste time being stuck for ages on one question, especially if it is only worth one or two marks.
- ✓ **Return later** … in the exam to the question you are stuck on – you'll be surprised how often inspiration will strike!
- ✓ **Think positive!** You are well prepared, believe in yourself!

Good luck!!

Answers

Here you will find the answers to the Target your revision exercises, Exam-style questions and Review questions in the book. Full worked solutions for all of these are available online at **www.hoddereducation.co.uk/ MRNOCRALApplied**. Note that answers for the Test yourself multiple-choice questions are available online only.

STATISTICS

Target your revision (Statistics) (pages 1 –5)

1   i   The 100 people interviewed.

    ii   People aged 60 and over.

2   No because each person is not equally likely to be chosen.

3   Quota sampling

4   Two possible sources of bias, e.g.
    - Only people with internet access will be sampled.
    - Only people who follow the researcher on twitter (and their contacts) will be sampled.

    Other possible sources of bias are given in the online answers.

5   Option C

6   Continuous

7   This is a 'show that' question. Go online for a full worked solution.

8   32

9   Two distinct correct comments, e.g.
    - There are no cars with CO_2 emissions below $80\,g\,km^{-1}$.
    - There seem to be two groups of cars – one group with lower emissions and another group with higher emissions.

    There are more possible comments in the online answers.

10   17

11   Two correct comments comparing proportions (not amounts) of electricity from different sources, e.g.
    - In December 2017 a lower proportion of electricity was generated using coal.
    - In December 2017 a higher proportion of electricity was generated using natural gas.

    Other possible comments are given in the online answers.

12   i   14.95 cm

     ii   1.15 cm

13   Two distinct correct comments, e.g.
    - Hybrid cars tends to have the lowest CO_2 emissions.
    - Petrol cars tend to have the highest CO_2 emissions.
    - Diesel cars have the lowest spread in CO_2 emissions.

    There are more comments in the online answers.

14   360

15   Two distinct ways to improve the presentation of the data, e.g.
    - Do not use a black background.
    - Do not use a 3–D effect.
    - Have a grid to allow reading values off.

    Other possible improvements are included in the online answers.

16   38.5 years, assuming that the first group is 20–25 and the last is 60–65.

17   10.6 years

18   Go online for a full worked solution.

19   i   0.59049

     ii   As long as the bars with vouchers are well mixed with the bars without vouchers, it is reasonable.

20   $\dfrac{5}{18}$

21   0.6

22   Not independent with correct working (see online answers).

23   $\dfrac{2}{3}$

24   $\dfrac{7}{41}$

25   Not binomial with justification (see online answers).

26   0.0302

27   9.2

28   Neil is wrong (see online answers).

29   $H_0 : p = 0.5$ where p is the proportion of male
     $H_1 : p > 0.5$ students in the college who like football.

30   $X \geqslant 25$

31   There is sufficient evidence at the 5% level that more than 30% of adults have high blood pressure.

32 i 0.2

ii 0.7

33 The Normal model appears to be suitable because the histogram is approximately symmetrical and bell–shaped.

34 0.0033

35 $\bar{X} \sim \mathrm{N}\left(37.4, \dfrac{2.8^2}{10}\right)$

36 $H_0 : \mu = 0.2$

$H_1 : \mu > 0.2$

where μ is the mean reaction time in seconds for the population of tired adults.

37 The p–value is 0.0004 (to 4 s.f.)

38 See online answers.

39 Two distinct correct comments, e.g.
- There is no correlation between number of teachers and pupil–teacher ratios.
- The pupil–teacher ratio for larger schools is fairly consistent at around 15.

Further comments are shown in the online answers.

40 There is sufficient evidence at the 1% level to suggest that there is positive correlation between cycling time and running time for triathletes.

Data collection (page 11)

i All customers of the supermarket.

ii There is no list of all customers.

iii Quota sample.

iv One possible improvement. Two possible examples are given below.
- Increase sample size.
- Sample at more than one time of day.

Interpreting graphs for one variable (page 18)

i The first chart is not appropriate because adding percentages for different areas is not meaningful.

ii The recycling rate in each area has gone up over the time period. In each area, the rate has stopped increasing.

Other answers include: recycling rates in London are lower than in the other two areas.

iii The chart shows that about 30% of household waste in London is recycled. This is less than a third but it does not refer to the percentage of households. If less than a third of households recycled their waste, they would need to recycle everything while the other households recycled nothing.

Averages (page 23)

i 7.5 is not correct; this is half way between 0 and 15 but the oldest person in the group could be nearly 16 so the midpoint is $\dfrac{0+16}{2} = 8$.

ii See full worked solutions.

iii Britney's answer is only an estimate because the data were grouped. The exact ages are not known and so the midpoint of each group has been used to estimate them. The National Statistics' value could have used the exact ages.

Diagrams for grouped data (page 32)

i For the blue graph, in the 20–30 group there are $40 - 12 = 28$ people.

For the red graph, there are $8 - 0 = 8$ people.

In the histogram, there are $10 \times 2.8 = 28$ people.

The blue graph must be the one representing Paper 1.

ii A From the histogram, it is $1 \times 10 = 10$.

B $136 - 94 = 42$

iii Overall, the students did better on Paper 2 than on Paper 1.

Medians and quartiles (page 40)

i The upper limit of the group's earnings.

ii A There are 100% of households altogether and half of this is 50%.

B £31 000 (value between £30 000 and £32 000).

iii Those with income over £87 000.

Standard deviation (page 45)

i The mean is 160.17 cm (2 d.p.).

The standard deviation is 6.38 cm (2 d.p.).

ii The mean is 159.97 cm (2 d.p.).

The standard deviation is 5.99 cm (2 d.p.)

iii The answers in part **i** are estimates because the data have been grouped and so they are less accurate than the values in part **ii**.

Working with probability (page 54)

i 0.968

ii 0.832

Conditional probability (page 57)

i $\dfrac{3}{18}$

ii $\dfrac{15}{28}$

iii $\dfrac{3}{13}$

Introducing the binomial distribution (page 63)

i   A 0.1413

   B 0.9576

ii For the binomial model used to be valid, the probability of students at the school being dyslexic needs to be 0.1, the same as in the general population. See online full worked solutions for another possible answer.

Cumulative probabilities and expectations (page 69)

i   The percentage of people working for that employer who speak Welsh may be different to 23%.

ii 0.9948

iii 0.0809

Introducing hypothesis testing using the binomial distribution (1–tail tests) (page 76)

i $H_0 : p = 0.9$

   $H_1 : p > 0.9$

   where p = the proportion of children aged 11–16 in the area who have mobile phones.

ii There is sufficient evidence, at the 5% level, to suggest that the proportion of children with mobile phones has increased.

iii No because the p–value is 0.05%

2–tail tests (page 81)

i $H_0 : p = 0.05$

   $H_1 : p \neq 0.05$

   where p = the proportion of faulty plates produced.

ii The critical region is $X \geqslant 6$.

iii There is no evidence, at the 5% level, that the proportion of faulty plates has changed.

iv 1.39%

Discrete random variables (page 87)

i $p = \dfrac{11}{64}$

ii For $X = 1$, the first random integer must be 4 and this has probability $\dfrac{1}{4}$.

iii Go online for solution.

The Normal distribution (page 98)

i   Mean is 3.617 minutes to 3 d.p.

   Standard deviation is 1.160 minutes to 3 d.p.

ii 0.2540

iii The calculation in part **ii** suggests that over a quarter of students would take between 4 and 5 minutes, so not a good model.

Hypothesis testing using the Normal distribution (page 107)

i $H_0 : \mu = 176$

   $H_1 : \mu < 176$

   where μ is the mean height, in cm, of all the soldiers the historian is studying.

ii There is sufficient evidence at the 5% level to suggest that the mean height of the soldiers the historian is studying is below 176 cm.

Bivariate data (page 117)

i   Positive correlation with an outlier.

ii A The student who got 53 marks on Paper 1 and 25 marks on Paper 2.

   B No because he/she did not do all of the second paper.

iii $H_0 : \rho = 0$

   $H_1 : \rho > 0$

   where ρ is the population correlation coefficient.

   $0.0056 < 5\%$ so there is sufficient evidence of positive correlation in the population.

iv There were more than 7 students in the second sample. See full worked solutions online for reason.

Review questions (Statistics) (pages 118–120)

1 i   Wesley is wrong with justification (see online answers).

   ii Possible target group with justification (see online answers).

2 i   50 seconds

   ii 51.83 seconds

   iii The histogram is fairly symmetrical so a Normal model might be suitable. The data suggest that a mean of 60 would not be suitable as both the mean and the median were about 50 – the fact that these values were about the same for the model confirms that a Normal model might be suitable.

3 Mean = 1.06

  Standard deviation = 0.968

4 i   Positive correlation.

   ii George is right with justification (see online answers).

5 i   0.7

   ii 0.4

   iii No with correct reason (see online answers).

6   **i** $\dfrac{17}{50}$ **ii** $\dfrac{19}{44}$

7   **i** 0.0125

   ii No with reason (see online answers).

8   **i** 7.1

   ii 0.2814

9   **i** 12.0%

   ii 0.723

10 The mean is 12.7 minutes and the standard deviation is 1.6 minutes.

11 **i** $H_0 : \mu = 20$

   $H_1 : \mu \neq 20$

   where μ is the mean mass, in grams, of the population of biscuits produced.

   ii There is enough evidence, at the 5% significance level, that there has been a change in the mean mass of biscuits produced.

12 **i** H_0: There is no correlation between temperature and rainfall in the population.

   H_1: There is negative correlation between temperature and rainfall in the population.

   ii There is insufficient evidence, at the 5% level, that there is negative correlation between temperature and rainfall.

MECHANICS

Target your revision (Mechanics) (pages 121–124)

1   **i** 50 s

   ii

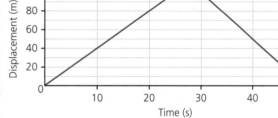

2 $3.3\,\mathrm{m\,s^{-1}}$

3   **i** $0.6\,\mathrm{m\,s^{-2}}$, zero, $0.05\,\mathrm{m\,s^{-2}}$

   ii 400 m

4   **i** $-1.6\,\mathrm{m\,s^{-2}}$

   ii 45 m

5   **i** 0.553 s

   ii $5.42\,\mathrm{m\,s^{-1}}$

6   **i** 13 m, 157.4⁰

   ii $\begin{pmatrix} 3.90 \\ 2.25 \end{pmatrix}\mathrm{m\,s^{-1}}$

7   **i** $\mathbf{a} = -2.5\mathbf{i} + \mathbf{j}$

   ii $\mathbf{s} = 16\mathbf{i} - 12\mathbf{j}$

8   **i** $x = 22.6t, \ y = 8.21t - 4.9t^2$

   ii 1.58 m, does not go over fence

9   **i** 3.78 m

   ii 21.6 m

10 $y = 1.5 - 0.1x^2$

11 $5g = 49\,\mathrm{N}$ vertically upwards

12

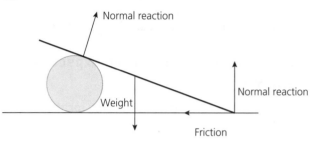

13 29.4 N

14 $x = -5, \ y = 27.2$

15 **i** 19 600 N

   ii $1.7\,\mathrm{m\,s^{-2}}$

16 6.05 N

17 Horizontally to the right for the 3 kg block:
$T_2 - T_1 = 3a$.

   Vertically down for the 1.8 kg particle:
$1.8g - T_2 = 1.8a$.

   Vertically up for the 0.5 kg particle:
$T_1 - 0.6g = 0.6a$.

18 $\mathbf{W} = -29.4\mathbf{j}, \ \mathbf{F} = 8\mathbf{i} + 6\mathbf{j}$

19 $1.5\mathbf{i} - 2\mathbf{j}$

20 $0.75\mathbf{i} + 4.4\mathbf{j}$

21 **i** $-5.18\mathbf{i} + 19.3\mathbf{j}$

   ii $W\sin 40°\mathbf{i} - W\cos 40°\mathbf{j}$

22 **i** 0.33 N is very small compared to the forces involved.

   ii $0.611\,\mathrm{m\,s^{-2}}$

23 0.404 N

24 $0.363\,\mathrm{m\,s^{-2}}$

25 0.279

26 1.09 m

27 **i** 0.098 N, 0.147 N

   ii 0.49 N

28 $X = 1.5\,\mathrm{N}, \ Y = 5\,\mathrm{N}, \ d = 100\,\mathrm{cm}$

29 **i** Not constant, see full worked solutions.

   ii Yes, when $t = 7\,\mathrm{s}$

30 66 m

31 $\mathbf{v} = \begin{pmatrix} 0.6t \\ 0.6t^2 + 2 \end{pmatrix} \mathbf{a} = \begin{pmatrix} 0.6 \\ 1.2t \end{pmatrix}$ not constant

32 $\mathbf{v} = (0.6t^2 - 1)\mathbf{i} + (2 - 0.1t)\mathbf{j}$

   $\mathbf{r} = (0.02t^3 - t + 5)\mathbf{i} + (2t - 0.05t^2 - 1)\mathbf{j}$

33 $y = 0.1(x + 2)^3 - 2(x + 2)$

Using graphs to analyse motion (page 133)

i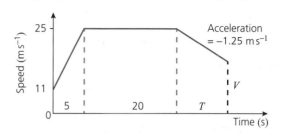

ii A $2.8\,\mathrm{m\,s^{-2}}$

    B $90\,\mathrm{m}$

iii $T = 12$, $V = 10$

iv A $21.6\,\mathrm{m\,s^{-1}}$

    B

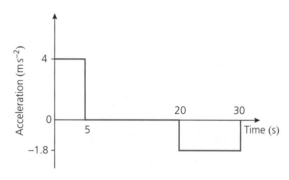

Using the constant acceleration formulae (page 137)

i $48\,\mathrm{m}$

ii $14\,\mathrm{s}$

Vertical motion under gravity (page 141)

i $40\,\mathrm{m}$

ii $27.5\,\mathrm{m}$

Motion using calculus (page 147)

i $a = 6 - 2t$

ii $t_1 = 1$, $t_2 = 5$

iii $\dfrac{32}{3}\,\mathrm{m}$

iv A Yes

    B No

v $15\frac{1}{3}\,\mathrm{m}$

Forces (page 153)

i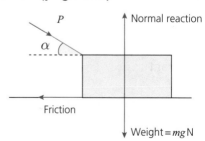

No, see full worked solutions for reasons.

ii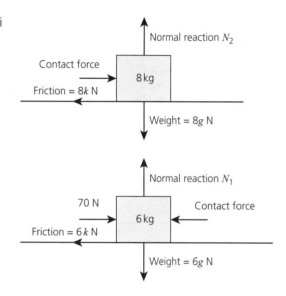

Newton's 1st law of motion (page 156)

i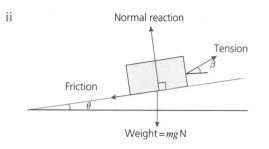

ii $40\,\mathrm{N}$

iii $58.8\,\mathrm{N}$

iv $30\,\mathrm{N}$

Applying Newton's 2nd law along a line (page 160)

i $1.5\,\mathrm{m\,s^{-2}}$

ii $5\,\mathrm{s}$

ii $28\,000\,\mathrm{N}$

Connected particles (page 163)

i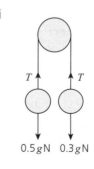

ii $2.45\,\mathrm{m\,s^{-2}}$

iii $3.675\,\mathrm{N}$

iv $4.9\,\mathrm{m\,s^{-1}}$

Kinematics in 2 dimensions (page 171)

i $(2\mathbf{i} + 2\mathbf{j})\,\text{m s}^{-2}$

ii   342° to the nearest degree.

iii never at rest

iv $\mathbf{r}_0 = 3\mathbf{i} - 2\mathbf{j}$, $\mathbf{r} = (t^2 - 4t + 3)\mathbf{i} + (t^2 - 2)\mathbf{j}$

v   1.25 s

Vector form of Newton's 2nd law (page 175)

i $-6.86\mathbf{j}$

ii $\mathbf{a} = 5$, $\mathbf{b} = 11.86$

iii $\mathbf{a} = -2.86\mathbf{i} + 1.63\mathbf{j}$

iv $\mathbf{s} = -35.75\mathbf{i} + 20.375\mathbf{j}$

Resolving forces (page 183)

i

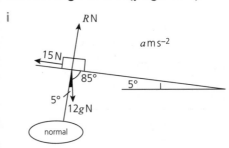

$a = -0.396\,\text{m s}^{-2}$

ii   2.84 m

iii 15.4 N

Calculus and vectors (page 187)

i   1.5

ii   18.1 m bearing 048.4°

Moments (page 193)

i   7.48 N at A, 6.24 N at B

ii   0.51 m from the handle end of the bat

Projectiles in flight (page 198)

i   This is a 'show that' question, so a short answer can't be given. Go online to see the full worked solution.

ii   H = 11.0 m to 3 s.f.

iii The ball travels 58.8 m before its first bounce.

iv   This is a 'show that' question, so a short answer can't be given. Go online to see the full worked solution.

Further projectiles (page 201)

i   15 m s^{-1} at 53.1° to the horizontal

ii $x = 9t$

iii $x = \dfrac{15}{2}\sqrt{2}\,t,\ y = 2 + \dfrac{15}{2}\sqrt{2}\,t - 4.9t^2$

iv   17.0° below the horizontal.

Friction (page 205)

i $2.64\,\text{m s}^{-2}$ ii   12.1 cm

Review questions (Mechanics) (pages 206–208)

1   i   30 m North   iii 30 s

    ii   90 m   iv   3 m s^{-1}

2   i   The particle is initially travelling north with a velocity of 20 m s^{-1}. It decelerates at a constant rate to 10 m s^{-1} after 10 s. It then travels at constant speed for the next 10 s. Then it decelerates, again at a constant rate, reaching 4 m s^{-1} after another 20 s.

    ii $-1\,\text{m s}^{-2}$, $0\,\text{m s}^{-2}$, $-0.3\,\text{m s}^{-2}$

    iii 590 m north of the origin.

3   5 minutes 9 seconds

4   2.81 s

5   i   See full worked solution.

    ii   10 m s^{-1}

6   i   2.42 m   ii   18.2 m   iii 48.8° to the sea

7   i   See full worked solutions.

    ii $D = 3.5\,\text{N}$, $N_1 = 78.4\,\text{N}$, $N_2 = 58.8\,\text{N}$, $T = 1.5\,\text{N}$

8   i   660.8 N   ii   548.8 N   iii 369.6 N

9   i   0.8 m s^{-2} ii   2160 N

    iii 340 N   iv   1232 N

10 i $\begin{pmatrix} 0.330 \\ 2.5 \end{pmatrix}\text{N}$

    ii $\begin{pmatrix} 0.165 \\ 1.25 \end{pmatrix}\text{m s}^{-2}$

    iii $\begin{pmatrix} 5.83 \\ 0.25 \end{pmatrix}\text{m s}^{-1}$

11 i   31.1 N   ii   66.8 N   iii 1.09 m s^{-2}

12 i   63.6 N   ii   0.200 (3 s.f.)

13 i   70 N, 0.171 m

    ii $F_A = F_C = 3$ N; F_A acts in the direction AB and F_C acts in the direction CD.

14 i $a = 4.2 - 1.2t$ At $t = 3.5$ he has reached his maximum velocity.

    ii   A 17.15 m

         B 61.25 m

    iii 14.77 s

15 i $\begin{pmatrix} 0 \\ 6t \end{pmatrix}$ not constant

    ii $\begin{pmatrix} 25 \\ 20 \end{pmatrix}$

    iii $y = (0.2x - 2)^3 - 0.2x - 2$

16 0.177 (3 s.f.)